"To Oindrila, whose passion for knowledge and unwavering curiosity illuminate the path for young minds everywhere. Your dedication in exploring the wonders of astrophysics not only makes understanding of the universe more accessible to young people, but also inspires countless others to dream big and reach for the stars.

We are proud of you and all young visionaries who inspire others to learn, innovate, and grow. May you continue to let your inquisitive mind lead the way towards boundless possibilities."

– Sarah Al Amiri, UAE Minister of Education, August 2024

Astrophysics

FOR YOUNG MINDS

Oindrila Maj

AF887711

Himalaya Publishing House

ISO 9001:2015 CERTIFIED

© **Author**

No part of this publication may be reproduced, stored in a retrieval system, or transmitted in any form or by any means, electronic, mechanical, photocopying, recording and/or otherwise without the prior written permission of the publisher.

All images, pictures, and figures given in the book are provided by the author

Cover Image by gene 1970, HUNG QUACH, OpenClipart-Vectors from Pixabay

First Edition : 2024

Published by	:	Mrs. Meena Pandey for **HIMALAYA PUBLISHING HOUSE PVT. LTD.,** "Ramdoot", Dr. Bhalerao Marg, Girgaon, Mumbai - 400 004. **Phone:** 022-23860170, 23863863; **Fax:** 022-23877178 **E-mail:** himpub@bharatmail.co.in; **Website:** www.himpub.com
Branch Offices	:	
New Delhi	:	"Pooja Apartments", 4-B, Murari Lal Street, Ansari Road, Darya Ganj, New Delhi - 110 002. Phone: 011-23270392, 23278631; Fax: 011-23256286
Nagpur	:	Kundanlal Chandak Industrial Estate, Ghat Road, Nagpur - 440 018. Phone: 0712-2721215, 2721216
Bengaluru	:	Plot No. 91-33, 2nd Main Road, Seshadripuram, Behind Nataraja Theatre, Bengaluru - 560 020. Phone: 080-41138821; Mobile: 09379847017, 09379847005
Hyderabad	:	No. 3-4-184, Lingampally, Besides Raghavendra Swamy Matham, Kachiguda, Hyderabad - 500 027. Phone: 040-27560041, 27550139
Chennai	:	No. 34/44, Motilal Street, T. Nagar, Chennai - 600 017. Mobile: 09380460419
Pune	:	"Laksha" Apartment, First Floor, No. 527, Mehunpura, Shaniwarpeth (Near Prabhat Theatre), Pune - 411 030. Phone: 020-24496323, 24496333; Mobile: 09370579333
Cuttack	:	Plot No. 5F-755/4, Sector-9, CDA Markat Nagar, Cuttack - 753 014, Odisha. Mobile: 09338746007
Kolkata	:	3, S.M. Bose Road, Near Gate No. 5, Agarpara Railway Station, North 24 Parganas, West Bengal - 700 109. Mobile: 09674536325
DTP by	:	Rajani Jadhav
Printed at	:	SAP Print Solutions Pvt. Ltd., Mumbai. On behalf of HPH.

This book is dedicated to my parents
(Sagarika and Chitro Majumdar).

I draw immense inspiration from Her Excellency Sarah Al-Amiri (UAE Minister of Education), whose remarkable achievements have sparked my passion.

The realization of this book wouldn't have been possible without the encouragement and exemplary influence of many individuals in my life.

Aunt Rita De Roeck,
Professor Freddy Delbaen
(Emeritus Professor of Mathematics, ETH-Zürich, Switzerland);
Professor Anja C. Andersen
(Astrophysicist, Niels Bohr Institute, Copenhagen, Denmark);
Uncle, Sergio Scandizzo
(European Investment Bank, Luxembourg);
Grandparents, Niva & Chhabiranjan Majumdar
Uncle, Chinmoy Majumdar.

In loving memory of my late maternal grandfather Debal Kumar Gupta whose legacy continues to inspire and shape my journey.

Preface

Our daily lives are caught up with the relatively mundane chores of living – this could include studying, passing exams or earning a livelihood. Every once a while, if we have allowed ourselves a moment of leisure, and we may have spent this moment, under a starry sky, chances are, our minds would have been filled with wonder, at what our eyes saw. The cosmos is truly magnificent sight to behold. Perhaps, a few of us might be wondered about the stars above, what forces made them, what laws held them in their orbits and what happened when they had reached the end of their existence. Certainly, many of us may have also wondered: Are we alone in the universe? Are there other solar systems like ours? How vast is the cosmos?

Our universe is literally everything. It includes all space and all matter that our cosmos contains. It consists of stars, planets, and other forms of matter and energy.

Scientists gave us the first knowledge and tools, and thus began a grand adventure of understanding the universe. Yet, our universe remains one of the biggest mysteries.

However, one doesn't have to be a scientist to contemplate mysteries. All it takes is a sense of wonder and some curiosity. Unfortunately, our education systems typically don't do much to cultivate this sense of wonder or curiosity. There are many books and textbooks that help students prepare for examination. However, there is a great dearth of books that engage and inspire a desire for this knowledge in reader's mind. This book is intended to fill this void.

This book delves into the wonders of our cosmos, serving as an ideal starting point for anyone, regardless of age, eager to embark on a journey into the vast unknown. Presented in simple terms, it aims to captivate students, those deeply curious about astronomy, and anyone seeking a clear understanding of cosmic phenomena, from particles like atoms, and molecules, to forces such as gravity, electromagnetic forces or waves such as X-rays, radio waves, and other natural and essential phenomena. This involves making use of infra red light from the electromagnetic spectrum and the James Webb Space Telescope (JWST) to look deeper into the space, enabling the observation of older stars and galaxies in our universe.

The book dwells on Einstein's theories, his equations and formulas, especially his Special Theory of Relativity that threw up the interesting questions such as, 'What is time'? Or 'Is time the same everywhere in space'?

We will also explore the main theory propounded by Albert Einstein, called the General Theory of Relativity, and examine how it influences our contemporary understanding of gravity today. We will be introduced not only to Einstein but also, preceding him, to Isaac Newton, another great luminary, whose Laws of Motion still guide spaceships today.

This book intends to introduce topics such as the Big Bang, the expansion of the universe and the intriguing subject of dark energy in a broader perspective, which can help enhance knowledge and interest in the field of astronomy. The expansion of the universe is a vast topic and includes many other topics which help us understand how the universe first began.

Scientists are still working on different theories and innovations to understand how our universe behaves and what might happen to our universe in the future. We can infer the existence or the presence of dark energy from how it affects earth and other astronomical structures or bodies, and how it causes galaxies to rotate or the expansion of the universe.

Tailored for middle and high schoolers, but accessible to all, this book aims to spark an interest in astronomy, presenting complex concepts in an engaging, story like manner. Written in easy-to-understand language, this book is an invaluable resource for youths and curious minds of any age, inviting everyone to unravel the universe's secrets.

Whether for educational pursuit or sheer curiosity, this astrophysics book welcomes all who seek to delve deeper into the mysteries of our cosmos.

Oindrila Majumdar

Contents

1

What is the Universe?

Our universe is everything. It includes all of space and all matter that our cosmos contains. It consists of stars, planets, and other forms of matter and energy in it. Galaxies, black holes, planets, stars, and other unknown objects and forces circling the universe waiting for them to be discovered. While we may have not yet fully uncovered the existence of our universe or its natural and dark sides, we have begun to invent and innovate new technologies that help us explore it more in-depth and reveal the hidden secrets beyond.

Our first adventure through the universe is how it all began, and the scientists helped us understand more and gave us the knowledge to uncover more vast topics about the universe. Our universe is one of the biggest mysteries that has left us unanswered through decades, many discoveries and observations have led us to uncover more about the universe and possible Alien life around other galaxies or planets. The cosmos is a vast place where forces and particles come together to create all the life forms, galaxies, black holes, and many more. Time and space are a crucial aspect of the universe, it shows how we view the universe in a three-dimensional perspective. Space-time was first proposed by Hermann Minkowski in 1908 to reconsider Einstein's equations and formulas in his special theory of relativity. Space-time also brings questions such as, what is time? Is time the same everywhere in space? Well, time is a whole chapter in itself but is probably the most interesting aspect of how we view the cosmos. It may seem that time is universal, that time is the same everywhere. But it isn't. In the special theory of relativity, Einstein disapproves of this theory and says, time is relative. It depends on where you are in the universe as time will slow down if you are next to a giant mass with a huge amount of gravity, such as a black hole. It appears that time would speed up or slow down around bodies that may have different masses and speeds. One of the many applications of Einstein's special theory of relativity is time dilation which is an example that shows how time can be shown relative to the observer. In simple terms, imagine you are on Earth with a clock, and there is someone, let's call them Rita, on a spacecraft out in space, also with the same clock at the same time. You are the observer from Earth, and currently, Rita is traveling at 90% of the speed of light relative to Earth or you. For you, 20 minutes have passed, but for Rita, only 10 minutes have elapsed, and time has slowed down for you. Many experiments were conducted to prove this time dilation. The experiment I just mentioned would not have been possible, as we cannot travel at the speed of light. Another experiment would involve a black hole, where you are on a spaceship and Rita is outside near the black hole. Time will slow down significantly for Rita as she is near an object with heavy mass and gravity, and as Rita gets closer to the horizon of the black hole, it will seem a lot slower for you as an observer. It will appear as if you are watching her in slow motion. Rita has already been sucked into the black hole, but for you, she is still contorting and distorting around the black hole and slowly she will disappear. The stronger the gravity, the slower time gets.

Photo Credit: ESO artist impression of massive object distorting space-time

Source: https://www.youtube.com/watch?v=Xi3hfw_7F1

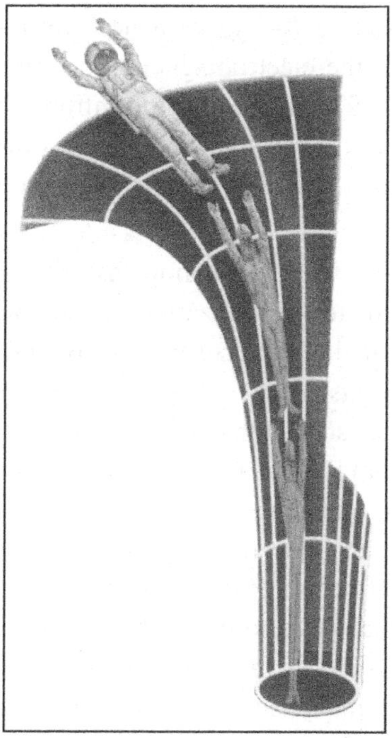

Photo Credit: Wikipedia

Source: https://en.wikipedia.org/wiki/Spaghettification

A natural phenomenon that has almost answered all our queries involves particles and forces. Almost everything around us is made of particles like atoms and molecules, and many are yet to be discovered. Forces are also essential in helping us move, creating interactions, and keeping us on the ground, such as gravity. Gravity is one of the weakest forces we know, but we also know gravity is a force that pulls us to the ground and attracts many other objects. There are many other forces such as electromagnetic spectrum or radiation which we apply on earth and scientists that have taught us the fundamentals of how things around us work. The electromagnetic spectrum was first predicted by James Clerk Maxwell in 1867. He said there should be different kinds of light that have longer wavelengths. In 1887, Heinrich Hertz proved and showed the existence of the waves Maxwell had predicted. Scientists also found it quite tricky to find shorter wavelengths in the electromagnetic spectrum but were later found. Soon after, different waves started being introduced, suc as X-rays as a new form of radiation, radio waves, etc. Through the electromagnetic spectrum, the James Webb Space Telescope (JWST) has been using infrared light to look further into space and look at the old stars and galaxies of our universe. With the pictures provided by JWST, we have observed something intriguing!

The universe has been expanding faster than we have expected and may also bring us to the realization of how old our universe may be. The expansion of the universe has always intrigued us, but how does it work? How fast does it expand? What is the cause of the expansion? There are so many vast questions and topics we will be discussing and learning about in this. The expansion of the universe is a very vast topic and includes many other topics which help us understand how the universe first began but what we do not know is that the universe has been expanding faster than we had anticipated. The expansion of the universe or the beginning of the universe first started with two particles or maybe more and which caused The Big Bang a theory that many scientists or people believe in how the universe is created. We can predict how other multiverses could behave or expand in the same way, we have realized that the universe is expanding due to dark energy, and we also know that because we have detected galaxies moving away from each other. For the universe to double its expansion, it would take more than a billion years, but the universe is expanding rather slowly in our time but if you see the average, it is going way faster. We do know the existence or the presence of dark energy because of how it affects us and different astronomical structures or bodies like how it causes galaxies to rotate or the expansion of the universe. This is a very vast topic that we must explore and we're still working on different theories and innovations to understand how our universe behaves and what might happen to our universe in the future.

So many scientists and space centres are operating missions to help us understand more about our universe's existence, and size, and to see the physical existence of cosmological structures that can also help in finding planets that we could habitat for the future. That's not the only reason we must find exoplanets, we must see and learn about the mysterious planets around us, how they work, behave, and how they get light or energy. This book is to introduce all these topics in a broader perspective which can help enhance your knowledge and interest in the field of astronomy. Now that we have a broader perspective of how this book may help in creating interest and helping you understand more about the depth of astrophysics before we get started, I wanted to add a few flavours of each of the chapters that are there to interest you and help you understand more about the topic.

GRAVITY

Gravity is a force that we all have experienced and have a perspective and understanding of how it works, how it behaves. But we do not know much about gravity, and we still have many observations and experiments to conduct to explore what else gravity can do or to provide a broader understanding of gravity and more. In the gravity chapter, we will discuss the latest theories and the main theory discovered by Albert Einstein that initially shaped how we see gravity today which is called the general theory of relativity, which we will know more about in the following chapters. As we know gravity is a force, a fundamental force that is the weakest of them all. Gravity has impacted or affected many areas of the universe like the cosmic web, the expansion of the universe, and maybe the beginning of the universe. We also have another type of gravity that also helps in the quantum world which is quantum gravity which we will explore more in further detail later. Quantum gravity shows the different aspects of gravity from a micro perspective. Gravity has been one of the first forces that have intrigued us in the field of science and knowing more about how the earth and beyond unlock many such discoveries and innovations. Gravity is a very important topic in the field of astronomy or astrophysics because of the contribution it has done to us in giving us curiosity and answers to our questions that have intrigued us for decades or centuries.

GRAVITATIONAL WAVES

Gravitational waves have intrigued us for years and we have finally discovered a physical gravitational wave that has surpassed our radars and interferometer thanks to the LIGO organisation. Gravitational waves are a significant part of the future, and it is involved in many topics such as motion of particles, the expansion of the universe, gravity, and many more. Gravitational waves go way back into history and beyond when the gravitational waves' first theory was created, and many such models were presented but back in the day people did not believe or did not wish to believe these such experiments or theories because they did not understand many sciences. This was because many mostly focused on religion and different beliefs but that is a whole different topic. More about gravitational waves and how today's science can help us innovate more technology to figure out more about the different structures or cosmological structures behaving in our vast universe and so on. The main reason we are still interested in gravitational waves is because gravitational waves have a potential and can help us figure out more about different whereabouts or gravitational waves always have a certain source like the one that LIGO had detected was from two rotating black holes, and we can figure out more about these supermassive black holes and different types of astronomical bodies. So, in this chapter, you will discuss lots about how the gravitational wave was discovered, why it was important, and how it can help us in the future.

LIGHT

Light is the key to our eyesight and to view the world we live in. We must all know that light is energy. Many beautiful sights, and creations can all be seen by light. Light shows our eyes magnificent moments and memories and even helps us realize our flaws and discomforts. Light comes along with everything as well as everything beyond our universe, it differentiates the possibilities and many things seen by the naked eye. Time travel can happen through the power of light using the infamous James Webb Space Telescope.

Light is not only important in the astronomy field but in biology, physics, math, and many other subjects, where light perspective is changed in every subject going from optics, nerves, our eyes, the speed of light, wavelengths, etc. One mission that widely used the true powers of light was the James Webb Space Telescope (JWST), which I mentioned earlier. It uses infrared light to capture stunning pictures of our universe in distances we never thought we could reach. It also uses a bit of time travel as well! As the light of distant stars takes billions of light years to reach us, we can see only how the stars looked when it was younger, who knows maybe the star has exploded or turned into a black hole. The JWST helps us investigate the early and young universe back when it was still evolving. Light can help us achieve and see things we thought we couldn't ever imagine but doing it in reality is hard. Reaching the speed of light could help us unlock a 'universe' of possibilities, this is how important light can be for us. A plethora number of scientists have helped us figure out light's properties and behaviours bit by bit and we still have a very long way to go.

THEORY OF EVERYTHING

In this theory of everything, we will discuss the different theories created in the past and today, that have helped us shape the idea that we have today from those theories and how it can help us in the future. It all started with the Pythagorean theorem where we had first gotten the maths and scientific knowledge to move on to the other theories such as general activity, string theory, Newtons law's, the uncertainty problem and so many more. These all help us in understanding the situations or astronomical bodies that we discovered and how they behave and how they might interfere with different planets or stars that might cause a revolutionary breakthrough. We will also go deep into history and how the different theories were created and how they explain today's scientific advancements with the help of the knowledge of many great scientists, physicists, mathematicians or astronomers, and many such who competed in the scientific field. All these theories come together to help us see the universe we believe today; many theories were made on how the universe began, and what was the reason? and many such questions that are still unanswered such as: Are we alone in the universe? Those theories in the past and theories we create today can help us answer today's questions and those questions that are still unanswered.

CONCLUSION

Astronomy is a subject that sparks interest in all but slowly loses that spark after a few years because of the lack of schools teaching and the exposure of many the interest in children in all subjects. For me, astronomy is the ultimate frontier where all the latest innovations are being sent to space and collecting valuable data which can be used later for much more future discoveries and more. Space is full of unique and knowledge-worthy information that can help us unlock the universe's deepest secrets hidden below, so join me on this quest to take a long journey into our mysterious cosmos through the stories and science in this book!

❄ ❄ ❄

2

What is Gravity?

Gravity has been around for centuries, and it has triggered our quest to understand the different forces of nature. However, we must first inquire: What is gravity? Gravity is one of the fundamental forces of the universe, a natural phenomenon that pulls objects to the ground, a force that attracts energy, mass, planets, stars, galaxies, and light!

The history of how the theory of gravity came into existence goes back to the year 1687 when it was discovered by Sir Isaac Newton. However, the idea of gravity was thought of, a long time ago, in ancient history. Back in ancient times, the Greek philosopher Archimedes presented the idea of the 'centre of gravity'. The theory suggested that if two equal weights, such as two basketballs that don't have the same mass, affect the centre of gravity of both balls then the two basketballs together should be equal to the gravitational pull, towards the centre of the earth. This was only one of the ideas on gravity. Another one was from an Indian astronomer Aryabhata, who questioned why objects didn't fall out while the whole of the earth spun around, while rotating.

THE LEANING TOWER OF PISA EXPERIMENT

Long before Isaac Newton there was Galileo, a great astronomer who also played a big role in proving the idea of gravity. Back in the years 1589 to 1592, Galileo performed an experiment that changed the idea of gravity forever, called "Galileo's Leaning Tower of Pisa experiment". In the experiment, from the top of a building called the Leaning Tower of Pisa, a slightly reclined structure based in Italy, Galileo dropped two spheres, with different masses. He dropped both at the same time, to see which one landed first. Which one do you think landed first? Surprisingly, both landed at the same time on the ground! Usually, we would think that an object with a higher mass would've landed first, because the heavier the object, the more acceleration it will experience (acceleration will be introduced in the further chapters).

How can two objects with unequal masses fall with the same acceleration, at the same time? Imagine two objects, one heavier and one lighter than each other, connected by a string, and which you from a tower; usually, the heavier mass would accelerate faster, and at the same time, the lighter object would be pulled by the heavier object. But in this experiment, the lighter objects are heavier than the objects with a higher mass, thus both objects fall faster and eventually fall at the same time while also accelerating at the same speed.

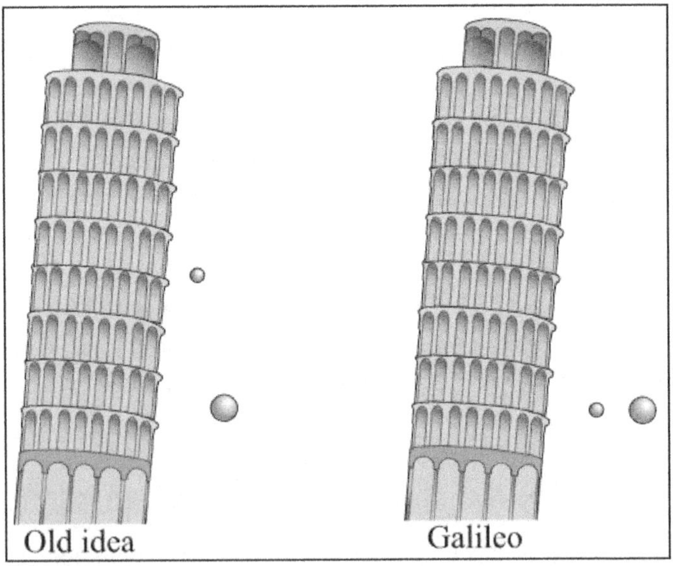

Photo Credit: Wikipedia

Source: https://en.wikipedia.org/wiki/Galileo%27s_Leaning_Tower_of_Pisa_experiment

Galileo's experiment disapproved Aristotle's theory of gravity, which states that objects fall at a speed equal to their mass. This caused a whole debate between Galileo and Aristotle on gravity. Galileo decided to do a physical experiment to prove his prediction and Aristotle was indeed proven wrong. Finally, a few decades later, based on these questions, observations, and experiments, Isaac Newton formulated his theory of gravity.

THE APPLE FALLS ON NEWTON!

Sir Isaac Newton was a great mathematician, astronomer, and physicist. He went to Trinity College in Cambridge, where he proposed the three laws of motion and the law of gravity which we will dig deeper into, in the coming pages. Gravity is one of the weakest forces in nature, this has been studied and proved by Newton with his mathematical formulas such as F = G [(m_1 × m_2)/r^2] which we will understand shortly. The story of how Newton came up with the idea of gravity has been around for decades.

Photo Credit: Adobe Stock

Source: https://stock.adobe.com/ae/search/images?k=newton+apple+tree &asset_id=497431948

According to the story, one day Isaac Newton decided to go out for a walk in the apple orchard. He found an apple tree, sat down, and started to think about the fundamental forces of nature. After a while, the sun began to set, and suddenly apple fell over Newton's head. He was startled and wondered if an invisible force caused the apple to fall? Why did it fall? Newton had so many questions, and he immediately ran to his study to learn more about this phenomenon. After studying and analyzing the theories and experiments done in ancient times, he went on to find more theories about gravity, and how it else it could be connected with the world of astronomy. He then thought of the question: How does the moon orbit the Earth? Now, we know that Earth's gravity attracts the moon and causes it to orbit around us, but back in the day proving scientific theories that could not be physically experimented with, was hard, because the majority wouldn't agree. An example of this is the Big Bang theory. Religious people, even in fairly recent decades refused to believe there was a start to the universe, because this contradicted their traditions and beliefs, so they banned all theories and even threatened the scientists with dire consequences. But today, the world has evolved.

Why the Earth Appears to Stand Still

However, the same thing happened when people refused to believe the earth rotated around the sun and many other planets too. They thought the Earth was the centre of our solar system, and before the idea of the heliocentric solar system was accepted, people believed that we were the only ones in the universe.

Newton's Laws of Motion

Back to the origin of gravity, when Newton got a clearer idea as to how gravity may work, he set out for the Trinity College in Cambridge, where he created the three laws of motion that can be applied to the motion of an object, and the force acting upon it. The first law of motion is the Law of Inertia which states that if a body is at rest or moving at a constant speed, in a straight line, it will remain at rest or keep moving in a straight line, at a constant speed, unless is it acted upon by a force. An example of the law of inertia is that if you roll a ball, it continues to roll unless a force or friction acts upon it and stops it, but the ball will always remain at rest. The law of inertia was originally formalized by Galileo Galilei in a Theory of

Horizontal Motion on Earth. Galileo had proved his hypothesis on the law of inertia, by rolling balls from inclined planes. However, Galileo reflected that when our earth spins why we do not feel the motion in space and he wanted to explain that. This is where the law of inertia comes into effect; while the earth is in constant motion, it appears to be at rest. But even if other forces, such as friction and air resistance, act upon us it will appear as if we are at a standstill.

Another real-life example that persuaded people that we were constantly moving was the sun and moon. When we discovered that one part of the earth faced the sun and the other faced the moon, that was enough proof that we were rotating and orbiting around the sun. However, some people still refused to believe it and stayed with the belief that the Earth was the centre of our solar system. The law of inertia has helped us understand the fundamentals of how gravity on earth and the universe works. Now, let's go into the 2nd law of motion. Newton's second law of motion states that the time of change of the momentum of a body is equal in both magnitude and direction to the force imposed on it. The momentum of the body is equal to the product of its mass (m) and its velocity (v). Let's break this down to further understand the concept of the 2nd law of motion. First, what is momentum? In short, momentum is the amount of motion that is acting upon a moving body or object that is measured by its mass and velocity or speed. Next, what is magnitude? Magnitude mostly refers to the amount of distance covered by the object while in motion or movement. Now that we have understood the two basic terms in the 2nd law, let's state it in simpler terms. Firstly, it states that the momentum (the motion of an object) of a body is equal to both the magnitude (quantity or distance of a body) and direction of the force acting upon it. Acceleration also plays a big role in this concept; acceleration (gaining speed) acts upon a mass or object when a force is imposed on it. An example of how this works is when a rocket is launched, its acceleration increases, and the rocket goes further into space, gathering in momentum and magnitude. The rocket has reached a certain distance because a force has been acting upon a moving object. The equation that explains it is $F = ma$, meaning Force is equal to mass times acceleration. The object, the diagram below will help you understand better.

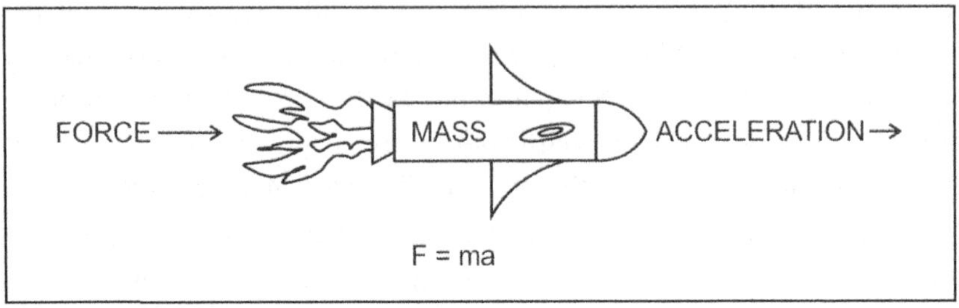

Photo Credit: Oindrila Majumdar

If a body has a force or net force acting on it, it is accelerating with it, but if there is no net force the acceleration will be 0. The 2nd law of motion has played a big role in the field of physics. It explains how forces can act upon different masses, which affect the acceleration of the object. The next law of motion is even more interesting. The 3rd law of motion states that "every action has an equal or opposite reaction", either a negative or positive reaction. An example is when you walk on pavement or a smooth pathway, you put a force when walking and as a result, you have a proportionate speed with little trouble. However, when you walk on sand you have trouble walking, why does this happen? That is where Newton's 3rd law is applicable, when you are walking and put force (the action) on sand, you get an opposite reaction or a negative reaction, meaning you don't get the same force you had put on the ground. Let's say we have a unit for this. You put a force of 5 units on a smooth ground and you will get the same force, 5 units. However, when you are on an unequal surface and put another force of 5 units, you do not receive the equal force. This law can also be applied to rocket launches! Now that we have covered all the 3 laws of motion, we can go into the main part upon which we built the idea of gravity, viz. the law of gravitation.

Newton wanted to understand how the moon revolves around the Earth and why the moon doesn't free fall and instead orbits the Earth, like it was attracted to Earth. Newton had assumed the existence of an attracting force between all giant bodies that caused them to be attracted to each other, and that the attractive force did not require contact between two bodies, but a force that could act at a distance. With the help of the law of inertia (the 1st law of motion), Newton concluded that a force from the earth was causing the moon to circulate the earth at a constant speed, rather than moving in a different direction, away from the earth, and that force was gravity, which he later discovered.

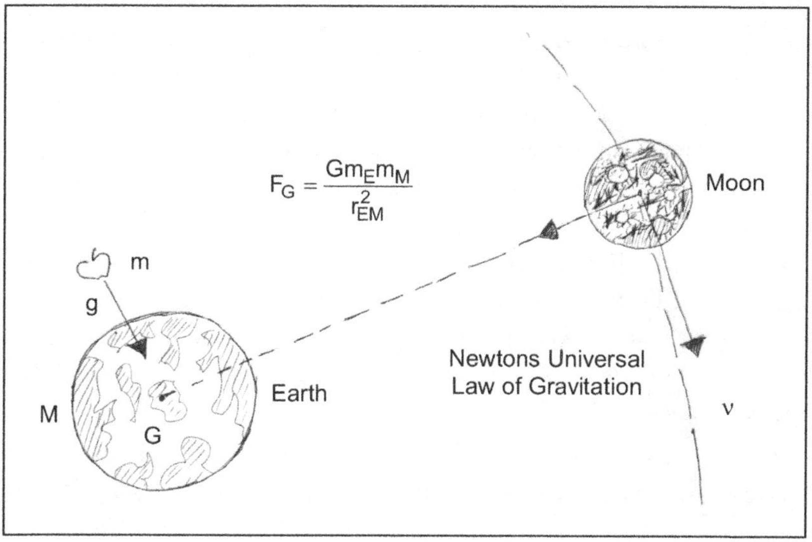

Photo Credit: Oindrila Majumdar

In Newton's theory, every particle of matter attracts other particles with a gravitational pull, meaning everything is attracted together by particles somehow. Depending on the strength of the gravitational pull exerted by a body, the more objects or bodies will be pulled towards it.

Here's a scenario: there's an asteroid heading close to our solar system. Now, just imagine that Jupiter, Earth, and Mercury are together when the asteroid is headed toward us. Which planet will it most likely be attracted to? The answer is, that it will fall into the orbit of Jupiter, why? The higher the mass, the higher the gravitational pull and attraction towards the asteroid, and thus the asteroid will be immediately pulled into Jupiter's trajectory and orbit there. Now that we have covered Newton's perspective on gravity, let us move to one of the smartest and most famous scientists of all, Albert Einstein.

Enter, Albert Einstein!

Albert Einstein discovered one of the most knowledge-changing theories that changed the Newtonian perspective on gravity through his General Theory of Relativity. This theory describes how the universe is like a fabric of space and time, called space-time, wherein different bodies and planets warp space-time and cause the time to slow down, as gravity slows down in certain areas. The idea of a fabric of space-time from the general theory of relativity discovered by Albert Einstein in 1905, changed the whole perspective of how bodies in space work and how they can curve space-time. The curvatures are all concerning time, how much time it may take for it to curve the frequencies of the waves, and maybe the speed of these curvatures. Later, Albert Einstein also developed a modernized version of the general theory of relativity, which he called The Special theory of Relativity.

So, we can say that in our solar system, our sun warps the centre of the solar system causing our planets to warp around it as well, whilst keeping a distance away from the sun. Now, let's imagine a surface that has a centre, which represents the warping of a body which also means that there's a small hole inside that curvature. Now, let's place a small ball inside that hole and imagine that it's our Sun. Now, let's place another ball around the surface which slowly curves and rolls into the hole, so it acts like a funnel, and when the ball is a planet going around the "sun" it starts to slowly move around it in an elliptical orbit. In this example, our earth ought to eventually fall into the sun, but why doesn't that happen in real life in our solar system?

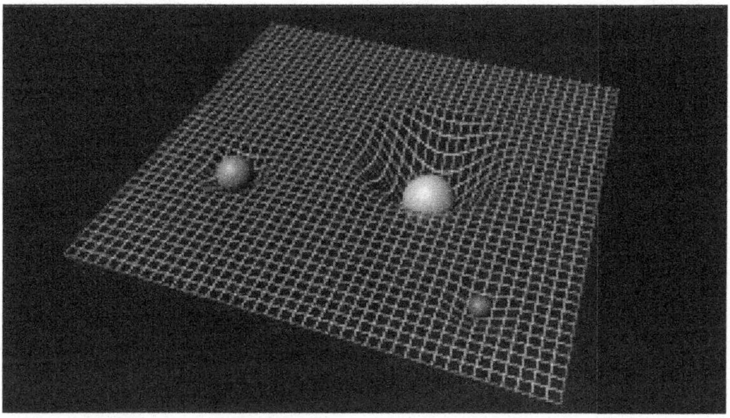

Photo Credit: The Conversation

Source: https://www.weforum.org/agenda/2017/10/stephen-hawkings-phd-thesis-explained-simply/

How Space-Time Fabric Works

Let us first understand how the space-time fabric works, how bodies warp around these space-time curvatures, and how they have helped us detect gravitational waves which we will soon learn in the next chapter. Did you know that The Pythagorean theorem, a vast topic in the field of mathematics, also relates to space time, and astronomy? The basic Pythagorean theorem is one of the first mathematical topics that were learned in the oldest civilizations, by Pythagoras, a Greek mathematician. But how does that relate to the space-time curvature or fabric space-time in our universe? let's imagine that in our space-time, there are three points of a Pythagorean triangle, let's say one point is delta, another one is delta squared, basically like different symbols, and let's say the space-time curves as it stretches. What might happen to three points and the structure of a triangle?

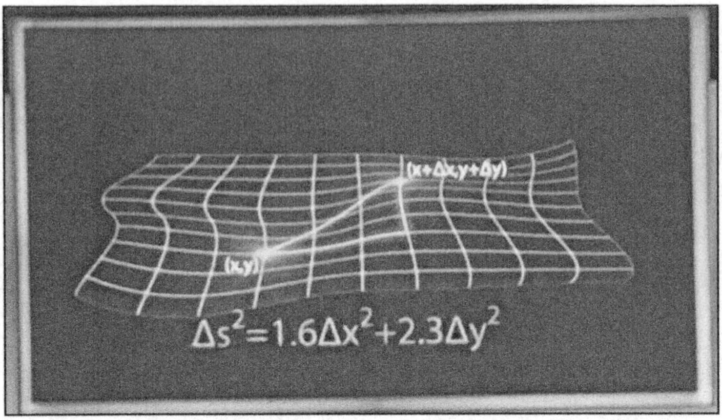

Photo Credit: Gravitational waves: A New Era of Astronomy/ World Science Festival

Source: https://www.youtube.com/watch?v=xj6vV3T4ok8 (36: 23 minutes)

If the fabric gets stretched, the triangle structure changes and the values of those points increase or decrease. This is called space-time geometry, where masses interact with the space-time fabric, by stretching and squeezing it, time by time. This can also relate to the 2nd law of motion $F = ma$. So, when our ball is the mass that accelerates, while in elliptical orbit, it means that a force is acting upon it, in this case, it's gravity. Now, because the higher mass of the "sun" has more gravity than the "planet", so the "sun" lures it in. But why aren't we getting closer and closer to our sun, while we have a systematic orbit around our sun? We will get to that later. Another aspect of the general theory of relativity is Euclidean geometry, which suggests that space is "flat", meaning it has no curvatures, but according to Einstein "flat" space only occurs when there is only space. When there is a mass or there are bodies, they cause space and time to curve or get distorted.

Why doesn't earth fall into the Sun?

Now back to the question of why we don't fall into the sun, if it has a bigger mass than us. The original theory is if the mass that moves in the curved or warped area is light, it must have less gravitational pull, and so it is pulled in by the higher mass. However, if it has enough mass to keep its distance away from the higher mass it is safe. If a higher mass had less gravity, then it could keep the rotating objects apart from each other.

Photo Credit: 5-MINUTE CRAFTS

Source: https://5minutecrafts.site/learn-world/why-the-earth-doesnt-fll-into-the-sun-1791/

Another theory could suggest that since our earth orbits around its axis, it can pull away from the sun, and when it is further away and its gravitational pull is acting on the sun. It is also avoided by other planets that also put their gravitational force on the sun, which causes both the pulls to be equal on both the planets and the sun. When space-time is curved, time slows down, and it curves and warps, which can be differentiated from the observer's perspective. The space-time fabric is three-dimensional i.e. it has length, breadth, and height. Now, let us compare different masses with different densities and see how much distortion is created. Let's start with our sun. The sun warps space-time lightly and slowly curves space-time. Our sun isn't as dense as a white dwarf, a white dwarf would warp space-time even more. Then there's a neutron star, it's even more dense than a white dwarf and is compressed as well, and makes space-time warp even deeper, which causes a neutron to almost have its singularity.

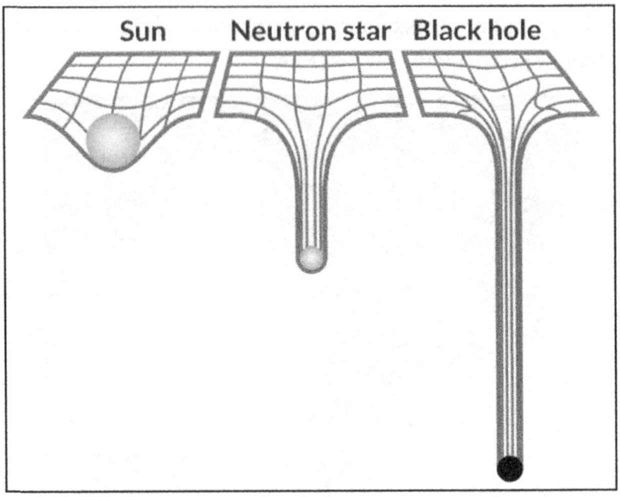

Photo Credit: Astronomyblog on Tumblr

Source: https://sfarnell.wordpress.com/2018/03/19/the-fabric-of-space-time-and-relativity/

What is a Black Hole?

However, out of all of these bodies, the most dense is a black hole, the last known observation of orbit is at the beginning of entering the black hole where time slows down for the observer. After that stage, the funnel gets deeper and you enter the real black hole into which you are sucked in, by the event horizon, which goes around the black hole and eventually, you reach the singularity. The furthest anybody has distorted is till the point of the event horizon. It is believed that a neutron star can reach below till the singularity, which is where the black hole distorts and the warp of time ends, it will slowly shrink and go beyond the Schwarzschild radius. What is the Schwarzschild Radius? The Schwarzschild Radius is also known as the event horizon of a black hole as we had talked about the warping of a black hole.

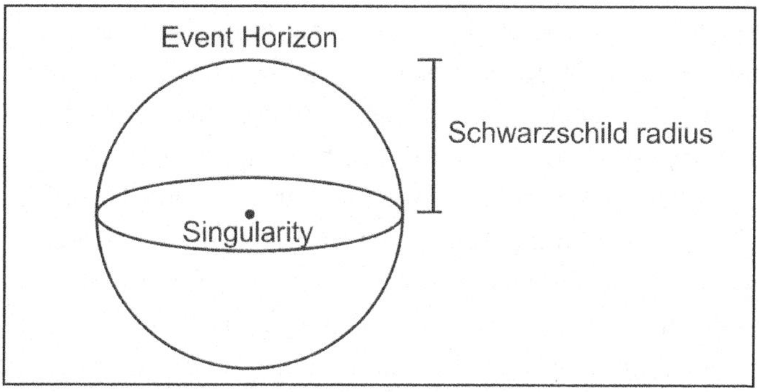

Photo Credit: Physics for Idiots

Source: https://physicsforidiots.com/space/black-holes/

When a black hole warps or distorts space-time it has four levels of warping- the first one, is the last stable orbit, where the body will immediately be sucked in by the black hole mass; the second is the black hole area where the original mass was located after which we have the event horizon where the Schwarzschild radius is reached; beyond that, the entire mass would be pulled out of singularity; the last stage of the black hole, singularity is a small, infinitely condensed point, that is the source of how black holes suck in most mass-less or mass objects- at this point information is lost. If the mass reaches beyond the event horizon or the Schwarzschild radius, space-time will be stretched to Infinity, which includes time also being stretched to Infinity, since space with respect to time is what the space-time fabric is made of.

Let's use the neutron star as an example- when it goes beyond the event horizon, a neutron star gains additional mass or weight, and it eventually collapses or shrinks beyond the Schwarzschild radius, and sooner or later mass could escape from its surface. This is only when space-time is distorted to its limits, but this is a very rare event, although the distortion of space-time through a neutron star is very high, a neutron star may eventually escape from its surface or the fabric of space-time. Now that we have covered the basics of general relativity, let us see the modernized version of a general theory that Einstein had created, viz. the special theory of relativity.

Before we get into that, a quick recap - what have we learned so far? Well, the Universe is a fabric of space and time, which when curved or warped, time follows that pattern and stretches and bends. But how does that affect a person's timeline? Let's imagine a rocket ship has two passengers, one passenger goes near a black hole, and the remaining passengers would begin to notice how time is much slower for them, than the time for the astronaut out there. This is all due to gravity, near a black hole; gravity is at its most intense which sucks in the astronaut, but observers can still see the astronaut descending into the black hole, which means them, gravity has slowed down time. A real-life example would be if you place two clocks, one on top of a mountain and the other under the sea, what would happen, would the two clocks show the same time? Well, let's think about gravity.

Photo Credit: A relativity cartoon (from npl.co.uk) illustrates gravitational time dilation

Source: http://www.leapsecond.com/great2016a/

The two clocks are at a distance, one is near the centre of the earth and one is far away from it. The centre of the earth has no gravity or G = 0 and if one of the clocks is on top of a mountain, gravity changes time and because of the distance it is also why they are different. Time has its different properties and one of them is gravity.

QUANTUM GRAVITY

Before understanding Quantum Gravity you must understand the term 'quantum'. Quantum comes from quantum mechanics, which is the study of small or micro-particles, and small energy packets called quanta. Quanta is also related to string theory, also known as the Theory of Everything, because it helps us discover new particles through combinations of vibrations represented on a string. You can find new combinations of particles with unusual vibrational patterns, that can lead to further discoveries initially atoms, particles, and more microscopic particles that make up our universe. String theory is also known as the theory of everything regarding the fact that particles are the ones that makeup almost everything in the universe. String theory will be later introduced in the following chapters.

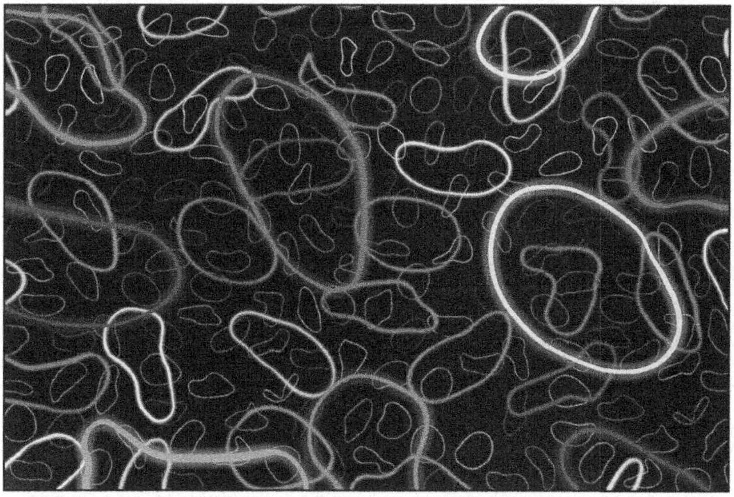

Photo Credit: Victor de Schwanberg/Alamy/ New Scientist

Source: https://www.newscientist.com/definition/string-theory/

There are three fundamental principles of QM (Quantum Mechanics) that shape the basics of how QM can be related to very small things, and how waves and light can be introduced in between.

The first principle is called 'Quantized properties', which says that certain properties, such as position, speed, and colour can occasionally occur in specific situations or amounts or combinations, that make a particle unique or different from all the rest. It challenges the fundamental assumption from classical mechanics, where all the said properties are the same for all particles, which constitutes a smooth and continuous spectrum. The 2nd principle is 'Particles of Light. This describes how light can behave as both a particle and wave. The theory that defines this idea is the 'wave-particle duality', discovered by Louis de Broglie. Moving on, the principle says that light can sometimes behave as a particle and other times as a wave, which was discovered long after the idea of light as a particle. Light can bounce off the surfaces, reflect, refract, and even bend! Waves can also act like this. There are two properties of a wave that are important in this principle, Crest and Trough. A crest is the top of a wave and how high it can get, a trough is the bottom of the wave or the downward side of a wave. What we are saying here is that when the crest is higher, it results in a brighter light, and when the trough is higher the candles go out and it gets darker. The third principle is 'Waves of Matter', where matter can 'theoretically' behave as a wave, which was discovered almost 30 years ago, showing that matter can exist like particles. We know that electrons define the atomic number of an atom, and we know that atomic particles can act just like waves, which brings us to the theory of wave-particle duality once again.

Bridging the quantum world and Gravity

Now let's focus on how Quantum is connected with gravity, and how they relate. Quantum mechanics is the study of small things, like particles gravity, and other forces, and something called the 'force particle', also known as a graviton. Graviton is a massless particle, which mediates the gravitational force that makes up gravity. However, this is just a theory that can be later proved by string theory.

There has already been a new particle that we think is a suitable particle to be present in force of gravity, the Gauge Boson. Another property in the subject of gravity is electromagnetism and magnetic forces. Electromagnetism is based on the connection between electricity and magnetism; an example of it could be a compass, generator, thermometer, etc. Another example could be when a magnet is stuck to a fridge because of a classical magnetic field or force, which causes it to attract to the fridge's metallic surface. However, if you go deeper into the magnetic field, there are quantum particles that create this magnetic field, including gravity. Electromagnetism, the strong and weak nuclear forces, together with gravitational force, are the 4 main forces of the universe. However, there are other theories and ideas on how the gravitational field comes in the field of magnetism and how particles behave inside them. The main question is how gravity meets the quantum world in this is a huge debate, which involves the general theory of relativity and quantum gravity. As we know, the general theory of relativity is on how space-time is a fabric that shapes the universe itself. However, if we zoom in closer, while space-time curves, there are microscopic particles that travel faster than light, ignoring the curves, while gravity is weaker on the space-time surface, so particles move carelessly throughout space and time. It is easier to find particles of bodies on the surface of space-time, but when you go to another level and try to find particles or try to measure the surface of space-time curvatures, the math becomes nearly impossible. Quantum mechanics also plays a big role in how space times curves, viz. when bodies distort or displace themselves into space-time, which slightly curves space-time; the second way is through gravitational waves, which we will go into depth about in the next chapter. Moving on, let's go back to how quantum mechanics microscopically plays a role in gravity and space-time curvatures.

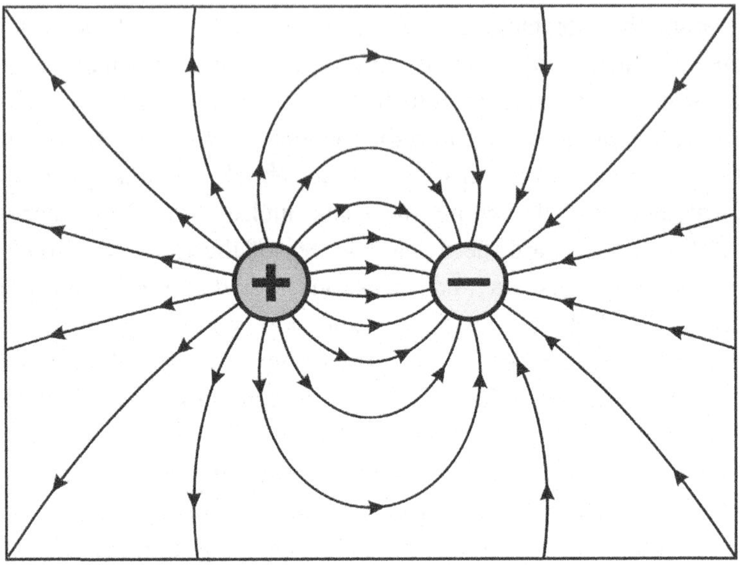

Photo Credit: Oindrila Majumdar

How Particles Behave

Back in the late 1940s, physics had created a new technique called re-normalization, for dealing with changes in the behaviours of particles, in the field of quantum mechanics; essentially, an electron boring trip around the nucleus can have infinite ways of orbiting the nucleus, and how these changes may affect the nucleus and other surroundings. A type of particle "waving" around the curvature of space-time called the graviton, a massless particle, with a small amount of gravitational field that pulls it to the fabric of space-time, but also lets it carelessly roam around the surface because it doesn't have much of a centre of gravity itself.

Different particles behave differently between the space-time fabric which also affects the distortion of particles and space-time. Particles also consist of gravitational waves that help us detect them here on Earth. This is the idea or theory of wave-particle duality, but how do particles behave in space?

Since there is no air resistance in space, particles tend to vibrate where particles cannot move, but send out frequencies of their small vibrating movements, which change how particles move in space. In the atmosphere, since air is a mixture of many gases, each particle interacts with each of the gases differently and has a different behaviour; there are four states of matter - liquid, solid, gas, and plasma - which effect how particles interact with matter. Another example of particles behaving and interacting with different energies is kinetic energy, where particles only interact with air resistance or in the presence of other gases. Kinetic energy is the energy associated with the motion of a particle. For example, a moving bicycle or a flight or clouds, anything in motion produces kinetic energy. Kinetic energy can be latent in all kinds of forms of energy. Kinetic energy depends on two things: the mass of the object and the speed of the object. Particles and kinetic energy will all be studied in the future following chapters on particle and motion.

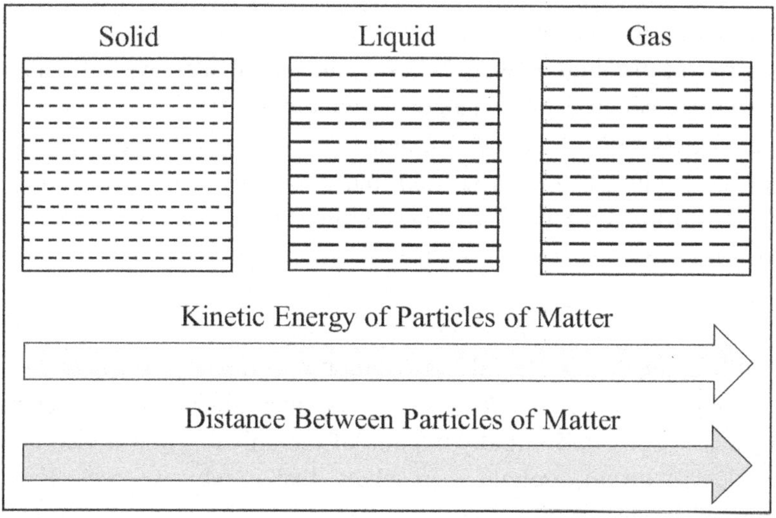

Photo Credit: Oindrila Majumdar

Our main goal is to understand how particles behave and interact with other particles and forces within the realm of space-time. This is why particles are a very important thing to learn, but how do particles come into gravity? We could try the idea of quantizing space-time into small, tiny particles, which have the force of gravity acting upon them.

We have learned a lot about the general theory of relativity and its clash with quantum gravity, let us get into the new topic of how gravity can also be a part of the idea of the expanding universe and how gravity may play a role in that as well.

THE EXPANDING UNIVERSE

The universe's expansion has helped us discover new bodies and galaxies that are being created even whilst the universe expands. However, researchers have found that the universe is expanding faster than we expected, but what is the cause of the universe's expansion? It's all because of dark energy, an invisible force that consists of particles that do not emit, reflect, and absorb light or interact with light which is why it appears invisible and is called dark, and which cannot be detected by electromagnetic radiation or any other high-tech tool. It may not be visible, but it shows its "power" by affecting the universe in many ways, such as the expansion of our universe or through how galaxies spin, and it also has effects on objects we can see or visualize. Even though dark matter may be invisible, its cause or effect has impacted the universe in such ways possible leading to the expansion and maybe even the beginning; it causes the universe to slowly accelerate, but it has been expanding faster than we expected. Another question that arises is 'does dark energy connect to gravity'? Dark energy does not interact with gravity, while it also makes up 68% of our universe and dark matter doesn't have any 'local gravitational effects' or any other gravitational force. Dark energy, as mentioned before, causes galaxies to spin, and initially expand of the universe and it also helps galaxies rotate and uphold themselves, that is why we see galaxies spinning due to the invisible force of dark energy and that is how we know the existence of dark energy. Dark energy mostly acts like the opposite of gravity because it exerts a negative energy towards gravity, which is not directly observed at the moment. Unlike normal energy or matter, dark energy cannot interact with electromagnetic forces or spectrums which makes it harder to see. Also, the particles that make up dark energy do not reflect or emit any light so we can only see the effects of dark energy like the universe's expansion, the rotation or spinning of galaxies and the motion of stars inside galaxies as mentioned before. We know that dark energy is the reason the universe is expanding which means that gravity doesn't mean it directly interacts with dark energy but both forces could contradict themselves as the universe continues to expand. Now that we know a bit about the role of gravity and dark energy in the universe, let us discuss the latest theories and discoveries on gravity on how that may help us to discover other forces that may act or behave like gravity, or forces that may act like the opposite of gravity.

The latest theory on gravity states that communication may be possible through space-time curvatures, but we know that sound is not possible in space, because there is no such source of air or water where particles of sound can travel through, so how will communication work in space? It may have something to do with quantum communication. Particles and electromagnetism waves are a part of quantum communication, a type of communication through different bodies and masses, but it can also be through signs or detection which can also be another way to communicate or reach and signal other people, same thing can happen in the situation through communicating in space.

Other than that, let's get into the new theories of gravity and how they may help in uncovering our universe in more depth. The latest theory on the gravity of our universe is based on two fundamental particles, with a few similar characteristics. This theory reconciles the two theories of quantum gravity and the general theory of relativity, which showcase the relationship between mass-gravity duality. These two theories are based on how different space particles behave in the presence of a body or mass on a warped or distorted surface. String theory resolves the clash of these two theories, by showing or creating new subatomic particles acting upon gravity or quantum gravity. Whereas light shows its wave-like properties that affect the particles that go at the speed of light and show the wave/particle duality. This shows how light can behave as a particle and as a wave which also can be applied to the subatomic particles' behaviour. In string theory, the string could go at the velocity of light and produce more particles. These particles have their characteristics which define or make them unique, and this helps us discover more particles in depth. So, these few binary characteristics are the different helictites of each particle, and how they behave. If they are similar, they may intertwine with each other, and this may create something new or redefine how we understand how string theory works.

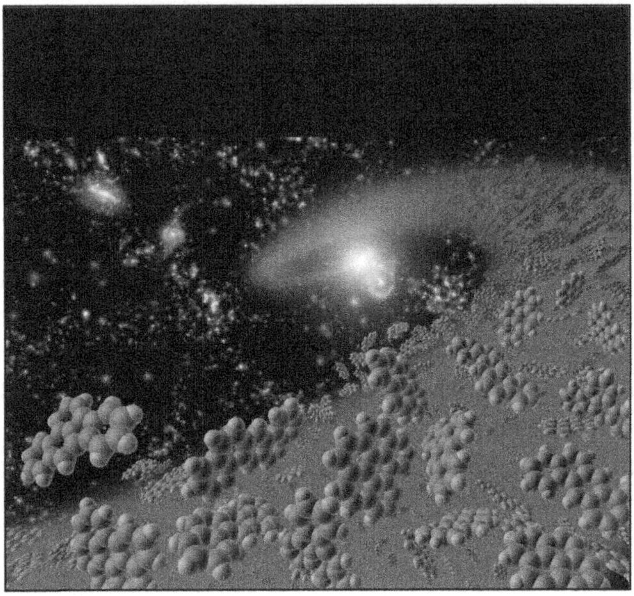

Photo Credit: NASA

Source: https://www.nasa.gov/image-article/ingredients-life/

These particles help in defining gravity's force, which contains a certain amount of energy that boosts gravity's force higher as well. This is due to the differences and elements in the particles that together act upon gravity, in the universe. The gravitational constant also plays a big role in this theory. The gravitational constant is where the gravitational force of two masses could be divided to find out the distance between them. This helps us in finding out the distance between the electromagnetic force between them which defines how much power or wave spectrum a particle has. String theory is based on the quantum reality of gravity and how the presence of particles in different forces can be affected in the universe, such as gravity. This new theory can help us unlock many aspects of gravity and its role in the universe, even though it clashes with many similar topics like quantum gravity, the expansion of the universe and many more to come. This theory focuses on mass-gravity duality, as mentioned before, or on how a mass particle can reconcile with its gravity as well.

Then there is the modernized or rechecked theory of the general theory of relativity, the special theory of relativity. The special theory of relativity is based on the physics of how speed affects mass. This can be related to how bodies rotate in the space-time fabric, maybe some could rotate at the speed of light. Two black holes rotating could be an example of this theory on how the speed of these masses affects the mass of the surface/fabric or even time. The rotation or speed of these masses affects the fabric which causes time to either distort or change in weird ways. Magnetic fields and gravity also play a role in affecting the body mass or speed that may or may not change or curve space-time, by its unusual behaviours. It goes through the electromagnetic spectrum that it can be detected, which helps us understand more about how special relativity works. This relates to the general theory more, because it affects the main part of the theory, the fabric of space-time, whilst this theory (special relativity) pertains to how bodies behave on the surface of the space-time fabric. These bodies rotate and produce a certain amount of velocity in the fabric which sends out gravitational waves which we will discuss in the other chapter.

The special theory of relativity talks about the electrodynamics of the speed of bodies. In the next chapter, we may talk about this theory in more depth, because it relates to more motion and waves. Now we have concluded this chapter where I will be summarizing all that we have talked about on gravity and other theories as well, as my opinion on how gravity may be able to help us understand other forces that could be discovered in the future. Gravity has already helped us in one way of seeing if certain exoplanets have a certain amount of gravity, which can be adaptable for us to live and sustain on. It has also helped in predicting how other galaxies or multiverses showcase their behavior in terms of gravity. Maybe gravity could help us in time travel or even deeper. Gravity has given us the knowledge to uncover many things we couldn't before. This gives us more curiosity to find more forces that behave or act the same or opposite to gravity, which may uncover more depth in our learning about the universe and beyond. Almost every mass or body has a gravitational centre, which is why understanding the main concepts of gravity is important in uncovering many different perspectives of Newton, Einstein, and many more brilliant scientists of this era's work, as well as the following generations ahead. Gravity helped us bring back particles, light, bodies, and many more to bring back the knowledge to apply to many other areas of the universe.

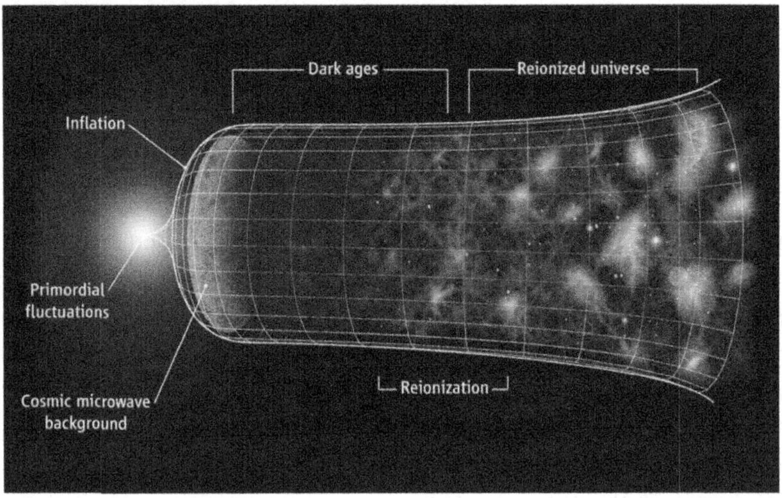

Photo Credit: C. Faucher-Giguère, A. Lidz, and L. Hernquist, Science 319, 5859 (47)

Source: https://www.forbes.com/sites/startswithabang/2019/02/19/this-is-why-we-arent-expanding-even-if-the-universe is/?sh=475883 da5311

❄ ❄ ❄

3

Gravitational Waves

Maybe, once when you throw a rock inside a pond and notice that after throwing the rock there is a pattern of circles formed in the water of the pond called ripples. These ripples inside the pond are waves sent out to other creatures - for example, when you throw some bread or food to fish, they sense the ripples and are alerted that something has been thrown, and immediately reach for the food. The same applies to the universe- but how can ripples be created in the universe? These ripples in space are called Gravitational waves.

HOW GRAVITATIONAL WAVES ARE CREATED?

High-mass bodies, such as merging black holes or neutron stars create gravitational waves in space, through a process called "gravitational radiation." When these massive objects orbit each other or collide, their immense gravitational forces cause ripples in the fabric of space-time. These ripples, known as gravitational waves, propagate outward at the speed of light, carrying energy away from the system, and causing detectable distortions in space-time, that we can measure with advanced instruments like LIGO and Virgo detectors on Earth. This phenomenon provides us with a unique tool to observe and study the universe's most energetic and cataclysmic events.

The first-ever gravitational wave was detected by LIGO (Laser Interferometer Gravitational-wave Observatory), an interferometer that detects gravitational waves through an electromagnetic spectrum. We will discuss this in further detail. Gravitational waves can also be expressed as disturbances in the space-time fabric, that not only cause 'ripples' which travel throughout the fabric, but even these ripples curve space-time at the same time. Gravitational waves can travel at the speed of light. The first gravitational wave detected may have been light years away, but it was situated in our galaxy which is why it was close enough to be reachable and detectable. This ability to detect gravitational waves can help us find more motions and movements happening in our galaxy and beyond.

Photo Credit: Photograph courtesy LIGO, T. Pyle

Source: https://www.nationalgeographic.com/science/article/gravitational-waves-ripples-space-time-ligo

GRAVITY AND GRAVITATIONAL WAVES

Gravity is implied in the process of gravitational waves. Gravitational waves are how gravity travels throughout the space-time fabric, while warping it in many areas. The speed of gravity or of gravitational waves is the speed of light which is how they reach Earth across large distances. Gravitational waves passing the earth produce a variety of stretching and pulling movements, but these happen only for a fraction of a second which makes it even more remarkable that we can catch them. Having passed by, the gravitational wave will slowly become weaker as it has already traveled billions of astronomical units. The frequencies and speeds of gravitational waves vary. Moreover, gravitational waves are based upon something called binaries, which show how gravitational waves, atoms or particles work inside. Does this make them very different from our usual particles, with different types of structures made from unknown sources, which in turn could help us in finding new particles? There is so much to learn about gravitational waves.

Photo Credit: Illustration: Olena Shmahalo

Source: https://nanograv.org/15yr/Summary/Interpret

Gravitational waves carry energy within the atoms and particles. This energy can be either positive or negative depending on the charge inside. Black holes make such motions that could potentially cause gravitational waves. We need movement and motion between two bodies to produce such waves. In laboratories, we are now trying to make artificial gravitational waves to see if this is possible.

The first gravitation wave is detected

Let's now explore how humans first discovered the phenomena of gravitational waves. We first, detected gravitational waves coming from two rotating black holes, which were causing a time of speed and motion between each other, in turn causing a warp or curve in the space-time fabric. This motion can create a distortion or disruption in the space-time fabric, which produces a signal which can take many light years or more to reach us.

The gravitational wave passes by Earth within a millisecond, which is why it is very hard to detect a wave. Luckily, the interferometer (LIGO) had been developed by then and it detected the gravitational wave, and that is how history was made. After this phenomenal discovery, many theories and innovations were devised to detect more such waves and to find out which black holes produced these waves. New laws and rules were created through this discovery, and Einstein was proved right as well!

GRAVITATION WAVES AND RELATIVITY

The first to formulate the concept of gravitational waves was Oliver Heaviside, in 1893. Oliver Heaviside, earlier had proposed the inverse analogy of gravitation and electricity. His idea was included in the Special Theory of Relativity discovered by Einstein, After Heaviside, it came to Henri Poincare, who modernized the idea more by rephrasing it as an emanating body that could travel at the speed of light while maintaining the Lorentz metric (which we shall cover in the following chapters ahead). By this time, people had put these two phenomena together, i.e. electricity and gravitational waves, making way for the term electromagnetic waves. Once Einstein had published his General Theory of Relativity, he was urged to apply it to Poincare's version, which proposed gravitational waves as 'gravitational dipoles'. Since then, many scientists have had many views, versions, and perspectives on gravitational waves, that all lead to the many theories such as, that gravitational waves could have different structures and wavelengths, dubbed as longitudinal–longitudinal, transverse–longitudinal, and transverse–transverse, which were all later then discovered by Hermann Weyl.

By now, scientists had figured out how gravitational waves travelled at the speed of light and measured them through a coordinate plane, to get an approximate idea of their speed and behavior. In 1936, Einstein and Nathan Rosen, an Israeli physicist, published a paper in 'Physical Review', a world-famous journal at the time, in which they said that the idea of gravitational waves could not fully be applied to the General Theory of Relativity, because of the influence of gravitational curvatures on the space-time fabric.

Photo Credit: Max Planck Institute for Radio Astronomy

Source: https://scitechdaily.com/einstein-proven-right-yet-again-theory-of-general-relativity-passes-a-range-of-precise-tests/

Gravitational curvature is a concept in Einstein's theory of General Relativity. According to this theory, massive objects like planets and stars cause a curvature in the fabric of space-time around them. In essence, gravitational curvature is applied to space-time fabric through Einstein's field equations. They mathematically describe how matter and energy warp space-time, and in turn, how objects move in response to this curvature. This theory has been confirmed through numerous experiments and observations, and is the foundation of our modern understanding of gravity.

However, when the paper was first reviewed by Howard P. Robertson, Einstein, as yet unfamiliar with this topic, with rage removed the manuscript from the 'Physical Review', never to publish there again. Later, Einstein's assistant Leopold Infield, went through the argument and convinced Einstein that the point Robertson made, was for a matter of fact correct, after which Einstein decided to accept it, and rewrote the paper with the correct idea of how gravitational waves worked, according to the Einstein general theory of relativity. In 1956, Felix Pirani, a British theoretical physicist, had a solution for this confusion as to how gravitational waves would work in the General Theory of Relativity, and he remade all the coordinate systems or planes, and added more knowledge on the singularity. He had now rephrased the name of Gravitational Waves to the Riemann Curvature Tensor, to be used in his work, which was done to solve the confusion between Einstein and Howard. At that time, Pirani's work was ignored as everyone's attention was on the question: 'Could gravitational waves transmit energy?' Richard Feynman, an American theoretical physicist known for forming the basis of quantum mechanics, brought up this question. He had proposed this idea, during the first 'GR' conference (the General Relativity conference where all idea of the theory was shared and discussed), at Chapel Hill, in 1957. Feynman showcased his work called the 'Sticky Bead Argument', which proposes to imagine a rod with beads in motion, producing heat due to friction of the beads along the rod, and which could be likened to the effect a gravitational wave would have upon passing through space-time fabric. Later, Hermann Bondi, an Australian-British mathematician and cosmologist, published a much more detailed version of the Sticky Bead Argument. The first evidence of the gravitational wave was discovered in 1974, by Russell Allen Hols and Joseph Hooten Taylor Junior, who experimented with the initial idea of a gravitational wave, through a binary pulsar. It is now called the Hulse-Taylor binary, which is a binary star system composed of different types of neutron stars and polestars. It is a highly magnetized rotating star that emits electromagnetic radiation, from different magnetic poles. This radiation can only be observed when a beam of light or emission is pointing toward the earth. This was how the first binary pulsar ever discovered in scientific history, and it was near an orbit that had a common centre of mass. This pulsar was then later used in different experiments and theories that helped develop real evidence on the gravitational waves. We can also compare these to radio emissions detected from only two neutron stars, which are phenomenally different orbital motions compared to these pulsars, and their centre of masses can be oriented towards the line of sight from Earth. The pulsar can also be used as a test in the General Theory of Relativity, discovered by Einstein.

Albert Einstein's Theory of Relativity talks about how things move when they're near something very dense and heavy planetary objects, like a planet or a star. It also explains how things like planets and satellites move around those planetary objects. Now, think of a star orbiting around a mass, and it starts sending out special waves called gravitational waves. These waves have energy, kind of like how the sun gives off light energy. Scientists use these waves to understand how things move in space. When scientists study pulsars (which are special stars), they learn even more about how all of this works. So, the theory of relativity helps us understand space and how things move, especially when we look at things, like pulsars and the energy they give off.

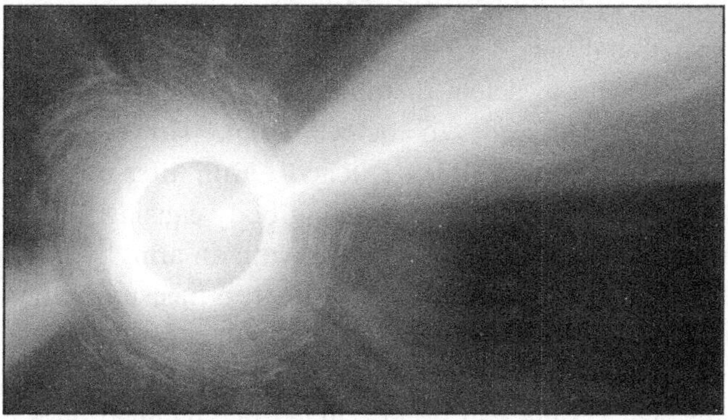

Photo Credit: Mark Garlick/Science Photo Library via Getty Images
Source: https://www.space.com/32661-pulsars.html

Now coming back to Russell Allen Holes and Joseph Hooten Taylor Junior who were awarded the Nobel Prize in physics, for the discovery of the first binary pulsar – a twin star system that provides a rare natural laboratory in which to test Albert Einstein's prediction that moving objects emit gravitational waves, as well as other aspects of his general theory of relativity.

Many scientists and physicists were awarded the Nobel Prize for adding to the knowledge about gravitational waves and how they behave in certain circumstances. As we know, the space-time fabric, according to the theory of general relativity, can act very differently; when space-time curves and distorts emitting different types of particles when it stretches, and when it squeezes, it creates different patterns of the gravitational wave.

Many were also awarded the Nobel Prize for physics for creating the technology that has helped us detect gravitational waves, in real motion. This can help future astronomical discoveries or in understanding many other types of gravitational waves (all gravitational waves are not the same), how gravitational waves might behave in a different multiverse, and how in a different multiverse, there is no space-time fabric, no type of distortion or curvature.

Now that we have covered most of the history of gravitational waves and what they were initially made of, let us get into the physics of gravitational waves.

The Physics of Gravitational Waves

Gravitational waves are constantly or always passing Earth, but we just don't seem to detect them, since they may have a frequency low enough for us not to detect them. However, when the gravitational wave GW 150914 reached Earth, after traveling over a billion light years, from another rotating body causing a gravitational wave ripple in space, LIGO detected it with an arm, by a thousandth of a width of a proton, which is equivalent to measuring the distance to our nearest star, outside the solar system. The effect of passing gravitational waves when extremely exaggerated or modeled in a coordinate plan system can be visualized or imagined on a perfectly flat region of space-time, with a group of motionless imaginary particles.

From the test model scientists are running, they intend to find out or imagine how gravitational waves may behave on the space-time fabric and how they might work in the future. The oscillation of the particles in the distortion of space-time, on the coordinate plane or model is exaggerated to show or represent a better view of how gravitational waves may occur again. The kind of oscillation the particle makes or is associated with, is the gravitational wave, which often has very small amplitude, as it is in a circular orbit, and is produced by a certain number of pairs of masses. This can also help us in verifying the gravitational wave size through the variations shown in the animation, and frequencies and spectrums they go through so that we can detect more of them, and predict how many waves or when a gravitational wave is passing towards the earth. This can help us figure out how gravitational waves are evolving and how different gravitational waves pass through the earth every single day or maybe every year. Gravitational waves may take billions of years or billions of light years to reach us.

Coming back to how gravitational waves behave, there are certain variations or spectrums that gravitational waves go through in terms of how they function and how they can be detected. Since gravitational waves have small amplitude, they need very sensitive detectors to observe and detect them. This is a reason why LIGO kept their technology and tools open-sourced so as to develop new technologies.

Every gravitational wave that passes through the earth comes in different ways from different sources of bodies that rotate or due to other reasons why gravitational waves occur. As mentioned before, gravitational waves go through a variety of wavelengths and frequencies. We must know the speed of these wavelengths and frequencies; for example, the animations that are shown on various online have a certain wave variation that they follow, which is very important to know in order to know how gravitational waves behave. Some gravitational waves may be small, but others could also reach about Earth's diameter or 47 times Earth's diameter. Gravitational waves are often polarised due to the nature of their source, and this polarisation at the source helps us understand how these gravitational waves occurred and how they behave in the space-time fabric. Another important part of gravitational waves is learning about the source of gravitational waves, and the patterns they may create on the space-time fabric. It all depends where it's coming from as when LIGO detected the first gravitational wave, they detected it from a rotating black hole.

In fundamental terms, gravitational waves are radiated or emitted by any body or mass with a certain motion or momentum, which involves acceleration or rotation, and which provides any kind of motion which is not perfectly spherically symmetrical. New research has brought up a hitherto unknown source for gravitational waves. When trying to detect the source of these gravitational waves, there was a slight disturbance that was interfering with the research; eventually it was found that there was a cocoon surrounding the black hole. For a while, this lead to the question if these cocoons were capable of producing gravitational waves, and whether LIGO was able to catch more of them and see them in a non-binary perspective.

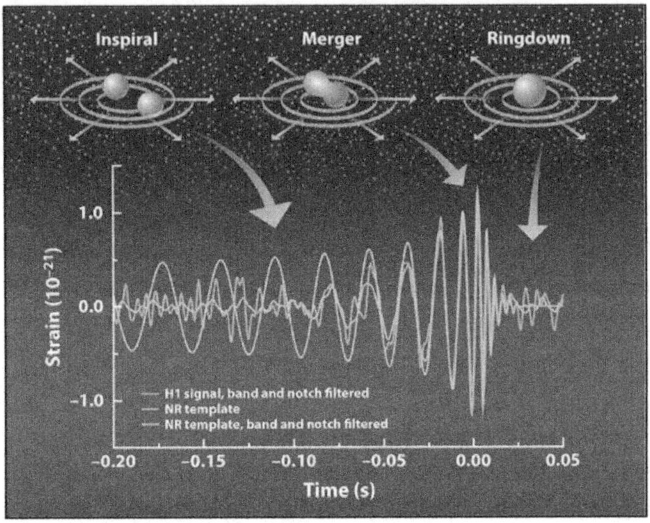

Photo Credit: Frans Pretorious, APS/Carin Cain

Source: https://physics.aps.org/articles/v9/52

Now an example that is commonly used is when a dumb bell or a spinning dumb bell is constantly in motion with the axis of symmetry or spherical symmetry, it will not emit or produce gravitational waves, but if it tumbles over the other end – in this case when two planets orbiting each other or rotating each other – then it will produce a gravitational wave, as how the first gravitational wave we had discovered by LIGO occurred. The heavier the dumb bell, the faster it tumbles, which means the gravitational wave would be of greater amplitude; in turn this means that it would give two gravitational waves from both planets sending a type of motion. In an unpredictable situation, when two of the weights of the dumbbell are two massive stars, either a black hole or neutron star, they would like orbiting each other very quickly; then large amounts of gravitational radiation will be given off.

Here a few detailed observations scientists have made throughout the gravitational wave process, as to how to detect the source of the gravitational radiation or gravitational wave. When two objects orbit or rotate around each other as a planet would orbit the sun, it would result in radiation or emission of gravitational radiation. Secondly, a spinning non-asymmetrical planetoid, let's say with a large bump on the equator, would also radiate as much as when a planet orbited the sun or two orbiting objects were rotating each other would radiate. Another example is that when a supernova radiates, it would mostly have a perfectly symmetric explosion while it radiates.

Another example is that of a spinning disc that can have angular momentum, and it will demonstrate the gravitational magnetic effect. One of the few exceptions is a black hole because black holes have a tremendous amount of gravity and density that distort space-time to its extreme limit. Moreover, black holes have an angular rotation momentum that helps in producing gravitational waves in a much more symmetrical way. As said earlier, in the dumb bell example, when it spins more, it gives a greater momentum which is how the ripples of the gravitational waves on space-time fabric occur. Another example is that of a supernova; a supernova does radiate when it explodes and has a high chance of radiating gravitational waves. Another could be a neutron star, or more specifically, a spinning neutron star which creates more momentum and angular moment of symmetry; this in turn radiates more gravitational waves and has a higher chance of rotating with another neutron star, as its density is way deeper into the distortion of space-time.

As we know, a mass distribution would emit gravitational radiation or gravitational waves where there is asymmetric motion among the bodies or masses. A spinning neutron star will generally not emit gravitational radiation because neutron stars are highly dense objects, and their mass distribution maintains a strong gravitational field that is almost perfectly spherical. Now, we must talk about the different binaries found in a gravitational wave, but first, what is a binary? Binaries represent a type of energy source for gravitational waves, involving orbiting bodies. An example could be a binary black hole, which can exhibit behaviour in the form of spiral, decreasing, or elliptical orbits. Another example could be a small and simple system of two masses in circular orbit, moving slowly, such as the earth and the sun. However, in this type of orbit, it represents a changing quadrupole moment. A quadrupole moment is one of the sequences of motions or charges, such as electric charge, current or gravitational mass, that can physically exist and be visualized. It is usually part of an expansion. This means a type of momentum or current that plays a role in the orbit of two masses or a binary. Binaries play a significant role in the occurrence of gravitational waves and their sources. Let us discuss the various properties and behaviours of gravitational waves; some of the types of properties include energy, momentum, and angular momentum, but there are many more beyond these few specific ones.

Water waves, sound waves, electromagnetic waves or light waves – indeed any type of waves that can carry energy, momentum and angular momentum. In doing so, they also carry momentum away from the source, that produces gravitational waves or gravitational radiation. So, for example, a binary system eventually loses angular momentum as the two orbiting objects rotate towards each other, through an angular momentum that produces radiation that radiates gravitational waves away from them. An oft repeated example is that of two super-massive black holes that rotate around each other, and likely emit a momentum that can produce a kick with an amplitude as large as 4000 kilometers per second, which is fast enough to eject a black hole completely from its Galaxy. Even if the kick is too small to eject the black hole completely, it can remove certain cosmological structures from the galaxy after which it will oscillate around the center, eventually coming to a stable pace and coming to a rest.

New research has been discovered on detected gravitational waves regarding the harmonious sound that the gravitational waves produce.

THE CONTRIBUTION OF LIGO

LIGO and its sister company are headed by the European Gravitational Observatory (EGO) called Virgo, which has a 3 km distance arm, at least 1 km shorter than LIGO's. These two help each other to lead the big discoveries and study gravitational waves, in depth. They have detected new sounds using radio sources, that they are now calling something like a chorus, with the help of pulsars which are stars that have a beat or a 'pulse', every time they rotate. Very recently, a major discovery came our way regarding gravitational waves. Low-frequency gravitational waves were discovered on Thursday, June 28, by the North American Nanohertz Observatory for Gravitational Waves (NANOGrav), marking a significant breakthrough after 15 years of research. The new gravitational wave detected was from two super massive black holes colliding 1.3 billion light years away, which had passed by Earth, and were luckily detected in a split second. Why must we detect more gravitational waves? What is the point of opening a whole new LIGO? How will it help science? Scientists have always relied on using electromagnetic (EM) radiation, such as visible light, X-rays, radio waves, and microwaves. Some people are also attempting to use neutrinos, which are subatomic particles. This information on gravitational waves provides scientists with a unique but extra perspective on the cosmos.

Learning all this new information on gravitational waves has given us a whole new perspective on how we view the universe and this has excited the whole science community. Throughout history, we have only theorized about gravitational waves, but we are extremely lucky to have state-of-the-art equipment to detect and hear gravitational waves. It is said that gravitational waves can travel at the speed of light. When gravitational waves travel, they stretch and compress the wavelength property of light, instantly! This may make it harder to detect, but we can identify new sources or maybe new ways that gravitational waves travel.

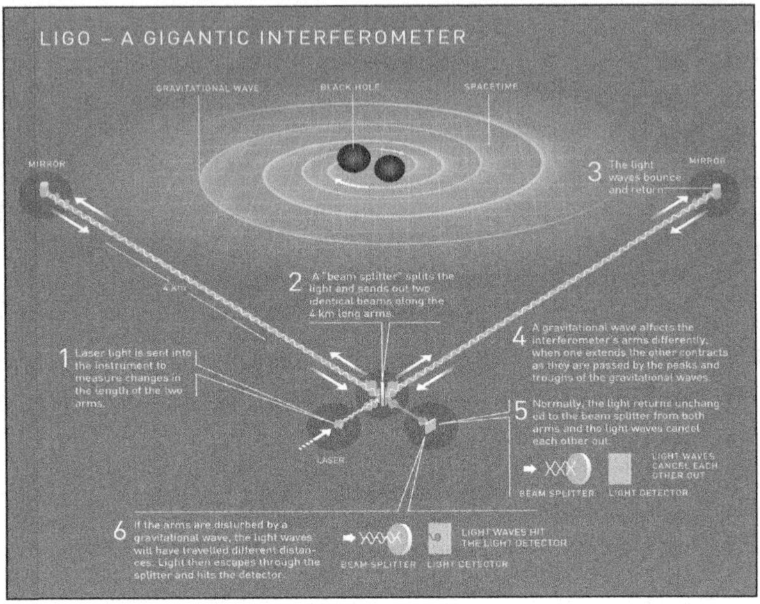

Photo Credit: Nobel Prize

Source: https://www.nbcnews.com/science/science-news/gravitational-wave-works-wins-physics-nobel-prize-n807081

❄ ❄ ❄

4

What is Light?

Chokher Aloy Dekhechilem Chokher Bahir-e

"My eyes once shed light on the world outside." by Rabindranath Tagore. Alternatively, we can also explicitly interpret this song of Tagore as follows:

With the sight of my eye

I looked for you outside of me Inside me I'll see

When light there is no more...

Photo Credit: WikiBio

Source: https://wikibio.in/rabindranath-tagore/#Nobel_Prize

BY THE GLORIOUS LIGHT!

Atoms, particles, grains, humans, animals, the world, planets, the universe, galaxies, stars, and many more, can all be seen and admired by one phenomenal force called light. First of all, we must all know that light is an energy. Many supernatural or unprecedented sights and creations can all be seen by light. Its reflection from our eye through the cornea, going deeper, light shows our eyes the magnificent moments of life and helps us analyze the universe, moments which have deeper connections with the brain and beyond it. Light plays with everything, as well as everything beyond our universe. It differentiates the possibilities and many things seen by the naked eye. One such possibility, time travel, can happen through the power of light.

Light changes how we see the world. Light is important not only in the field of astronomy, but also in biology, physics, mathematics, and many other subjects redefine light in different ways, which help us understand light way better than we could ever. As is said in the previous chapters, in different multiverses, light could act differently or maybe there won't even be something called light, or it could behave differently and have different particle mixtures and chemical bounds; different combinations that make each universe unique and we are genuinely stuck with the theory or phenomenon of light. Let's come back to the concept of light and then dig deeper into the more possibilities that light could give us, for future knowledge.

Photo Credit: NASA/JPL-Caltech
Source: https://www.space.com/mysteries-of-light-and-time.html

WHAT IS LIGHT?

In short, light is a type of electromagnetic radiation that can be detected by the human eye, and this radiation can occur over an extremely wide range of wavelengths. Here spectrums and frequencies can go up to almost less than 700 nanometers, which is how much the human eye can occupy, basically as much as we can see through the naked eye. Different types of light also exist, like red light which is down to about 400 nanometers, and then there's also ultraviolet light, which shares the same number of nanometers. The speed of light is a vacuum or a fundamental physical constant that is currently accepted as the value of exactly 299,792,458 meters per second, or about 186,282 miles per second.

There are many such ways in which light is displayed through the naked eye. Also, we measure things with light, as in light years and other speeds can be calculated when compared to the speed of light. All this is very important to learn about light. Having said that, there is no single answer to the question – 'What is light?', because light satisfies many constants in which light is discovered, known, and exploited. Physicists are interested in the physical properties of light, artists in an aesthetic sense of the visual world; our sense of sight on how light is a primary tool for looking at the fundamental world which helps in connecting minds and helping build a better future.

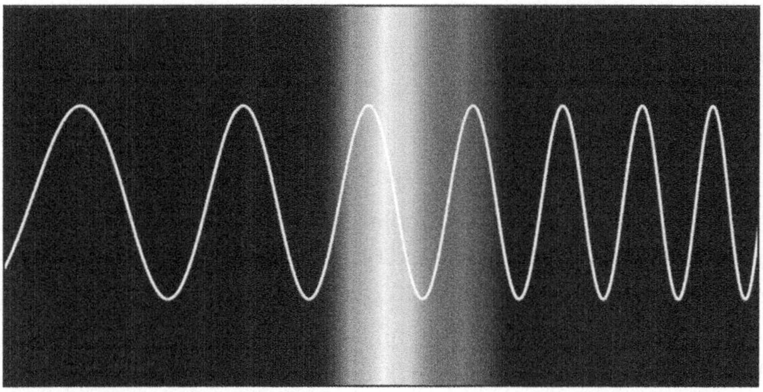

Photo Credit: NASA

Source: https://hubblesite.org/contents/articles/the-electromagnetic-spectrum

LIGHT FROM THE SUN AND STARS

When comes the broader aspect of light, which is the light from the sun which warms the earth, which drives global warming, and many other climate change problems that we have faced - all due to light. In a broader aspect, light can show us the universe; it is a way of time-traveling back to the beginning of The Big Bang and the universe. Recently the James Webb telescope was launched, which is way more advanced than the famous Hubble telescope. It will go deeper into the vast universe and reach out for a specific star beyond our galaxy, to see how stars and planets are structured beyond our own Milky Way. A secret about light is that light could be a way of time traveling, and this works by way of looking at stars or planets -the light that a star emits is its light. However, the light has taken many light years to reach us, depending on how far it is away from us. So, let's take a faraway star, for example. Its light may have taken a few million years to reach us. Thus, when we observe the star, we are essentially looking at it from millions of years ago. This is a phenomenon that helps us look into the depths of light, and how the universe was before. Looking this way helps in discovering more about the past or the origin of the universe, as well as studying and predicting the future of our universe. Before getting into the vast topics of light, let us dig deep into the history of how the discovery of light first began.

Humans Civilizations Understand Light

As we know, the discovery of light started way back when Thomas Edison invented the light bulb, which started the evolution of light. However, in the field of science and astronomy, the first detailed formula for the Wave Theory of Light was created by Christiaan Huygens, a Dutch mathematician–astronomer. He also brought in the laws of reflection and refraction and enhanced the idea of light to what we understand today.

The theories of light through history are a very important aspect of how we see and understand light today, and would understand in the future. It all started back in around 300 BC when Euclid wrote about optics, in which he had studied the properties of light itself. Euclid predicted that light travelled in straight lines and he described the laws of reflection and refraction from a mathematical perspective.

In 55 BC, Lucretius, a Roman who adhered to the theories and ideas of Greek atomists, wrote about how the heat of the sun provides the light we need for the Earth, which he called 'The light and heat of the sun'. He speculated that his experiment was composed of atoms and later knew about particle theory and how light can travel like a wave or particle. Another mathematician Ptolemy, had written on the refraction of light in his book 'Optics'. In India, many mathematicians, scientists, and astronomers developed the idea of light. In ancient India, the Hindu schools of Samkhya and Vaisheshika had developed theories for light. These schools gave a preview of the atomic theory of the physical world and showed how space and time are interconnected with it. They described the basic elements of the Earth like water, air, fire, and light and demonstrated how light can exhibit different aspects and characteristics. They also referred to sunlight as 'The seven rays of the sun'.

(Note that the equations and symbols on the image have nothing to do with space and time or light.)

Photo Credit: Del Hambre

Source: https://libraryforkids.com/greek-math-and-science-the-foundation-of-modern-knowledge/

After the Indian Buddhists developed a philosophy on the types of atoms that flash light and energy, they viewed light as equivalent to energy and that is how we know light as a type of electromagnetic energy. René Descartes was also part of the evolution of the theory of light. He had published a theory on the refraction of light and how light travelled faster in different mediums, and he also arrived at a conclusion on how sound travels through light waves. Descartes was incorrect about the speed of light in different mediums, but he was correct in assuming that light behaved like a wave. He also rightly concluded how refraction occurs in the speed of light in different mediums. Descartes was not the first to use these theories and ideas on light. Many were studying the mechanical properties of a luminous body going through different mediums. However, Descartes' theory of light was the start of how modern physical optics work today.

Now that we know a little bit about how light revolutionized our knowledge, we come next to the field of particle theory, which is part of the theory of light and introduces us to different angles of light and how these have brought us to the idea of light today.

During the 1660s, Pierre Gassendi, an atomist at the time, published a paper on the particle theory of light. Later, Isaac Newton studied this paper and developed his argument against the wave nature of light saying that light could bend and thus explaining the diffraction of light. Diffraction is not the opposite of refraction, rather it refers to when a wave interacts and encounters an obstacle or an opening; it is defined as the bending of waves, and since many people had believed that light can also behave as a wave, they also thought that light can bend, without knowing that waves can bend around corners or through a geometric shadow of the obstacle.

This diffraction has a lot of mathematical terms and complexities that can sometimes contradict some of the features of how light bends as a wave. There was an experiment done for diffraction, called the Slit Experiment. In it, you have a piece of paper with two slit openings and another surface behind it. When you shine a light on the paper, you see light going through the two slits, and this explains how light can be gathered in particles and not only waves. This experiment was done to prove the wave-particle duality theory by Louis de Broglie.

Coming back to Newton and his argument against this light bending through a wave; Newton's theory could only be used to predict the reflection of light, it could only explain how refraction works by assuming that light accelerated by entering a denser medium or that the gravitational force was stronger. Later, Newton published his final version of this theory in his Optics of 1704. His reputation helped his particle theory of light gain credence with other scientists and lay people, during the 18th century; he had erased other believable theories on how light can interact as a wave AND a particle.

Another mathematician Laplace, argued that if another body could be so massive and have so much gravitational pull that even light could not escape from it; we now know this to be a black hole. After wave theory was born, Robert Hooke, an English scientist, developed the "pulse theory"; he compared it with how light spreads through water, demonstrating a wave behavior. Then the mathematician Christiaan Huygens worked out a wave theory of light, in 1678, and published it later on. He proposed that light was emitted and reflected in all directions, while behaving as a wave through luminiferous aether (need help in explaining), on how the waves weren't affected by a gravitational force and it was said that gravity had slowed the waves down making it have a denser medium. Huygens wave theory predicted that light waves could interact with each other like sound waves do, to create a motion and sound that helps us hear through the air. Now, we shall look at some theories that helped us understand the technical view of light starting with the electromagnetic theory, by Michael Faraday, in 1815.

Faraday, an English scientist discovered the Inclined Plane Polarisation of Light, which explains how light rays travel along any magnetic field such as the space-time continuum. This was the first proof that light had something to do with electromagnetism. One year later, Huygens theorised that light might be a form of disturbance or disruption spreading along space-time. Faraday proposed another theory, in 1848, as to how light was a "high-frequency electromagnetic vibration" that could increase in the absence of another medium on the magnetic field such as ether.

THE QUANTUM NATURE OF LIGHT

Faraday's work inspired James Clerk Maxwell, a Scottish mathematician to study electromagnetic radiation and light. Maxwell had discovered that self-propagating electromagnetic waves can travel through space and time, at a constant speed which was measured by using the speed of light, as a form of electromagnetic radiation. He finally published his theory - A Treatise on Electricity and Magnetism. After the idea of electromagnetism was born, the quantum theory of light began. The quantum theory of light was first initiated by Albert Einstein who said that light is a composition of small packets filled with energy which are called photons, (Question: Isn't the small packets of energy called quanta?) which have wave-like properties. He also explained that the process of emitting electrons from metals, was like a strike of lightning, and this was called the photoelectric effect; an idea later led to our modern-day television and later to devices with screens. The quantum theory of light states that particles are like waves and waves are like particles; it also states that light acts like a wave and a particle. The quantum theory of light also talks about how matter and light behave like a particle and how both behave like a wave. The best theory that explains how light acts like both a wave and particle is Planck's quantum theory, as well as the wave-particle duality discovered by Louis de Broglie.

Let's start with Planck's quantum theory of light. Planck's theory explains a lot about the characteristics of light; light energy is emitted from a source in the form of energy packets that end out frequencies, called photons or it can be called quanta. Photons can act like waves, and also behave and interact with matter as particles, Planck also stated that an atom can only emit or absorb energy in discrete, small amounts, which relates to a particle's energy and frequency. Something else that Planck's constant introduces is black body radiation. Planck's constant explains how energy functions on the smallest sizes, such as with light particles called photons. It claims that a photon's energy is influenced by its color or frequency. Colors that are more vibrant and less vibrant both have more energy. Understanding how things function at the universe's most fundamental scales, such as in atoms and subatomic particles, is based on this idea. In simple terms, it creates a relationship between the energy of light and its color, helping us understand the behavior of the universe's smallest objects. Coming back to the topics, black body radiation is when a heated object glows and gives off a red-orange glow, once it's heated. Scientists and researchers have discovered that the wavelengths of light are emitted by the black body radiation; they also found out that the wavelengths found in the radiation can vary based on how much heat is applied. Energy, though originally quantized, can change in definite amounts, which is applicable to many different processes involving transformation of energy on the atomic or molecular scale in quantum mechanics.

WAVE-PARTICLE DUALITY

Wave-particle duality is the concept in quantum mechanics in which every particle can also be described as a wavelength, and in turn, waves behave like particles. The work of Max Planck, Albert Einstein, Louis de Broglie, Niels Bohr, and many other great minds have changed the way we see quantum mechanics today. The wave model for the theory was created by Louis de Broglie and was further developed by David Bohm. According to the hidden variable theory of quantum mechanics, particles have invisible properties that affect their behavior. In contrast to quantum mechanics' random perspective, it suggests certainty, in which particle properties determine outcomes. Experiments do not support the hidden variables' existence, however, and instead support the fundamental unpredictability of quantum events. But the phrase 'hidden variable' is not the correct saying since the variable stands for the positions of the particles so it gave more of the particle view of light but there was a more advanced model they had made on the wave point of view. In physics and chemistry, wave-particle duality states that light and matter inherit both properties of waves and particles, but the central concept of the quantum mechanics of light addresses the behavior of different quantum objects. The work of many great scientists mentioned earlier have all concluded that all objects have both wave and particle nature, in small scales like atoms. However, what is the significance of wave-particle duality? The importance of wave-particle duality is that it allows alternative equations to describe the real nature of light and matter, and how their wave functions usually arise, which is similar to the basis of quantum mechanics, and so all the behavior of light and matter can be understood.

The most common interpretation of a wave function is that there is a probability of finding a particle, at a given point, and these probability equations can interfere with other wave-like properties, as particles end up being distributed at the same time. The probability in the wave-particle duality is a very important aspect of this certain theory of how light can behave both like a wave and a particle.

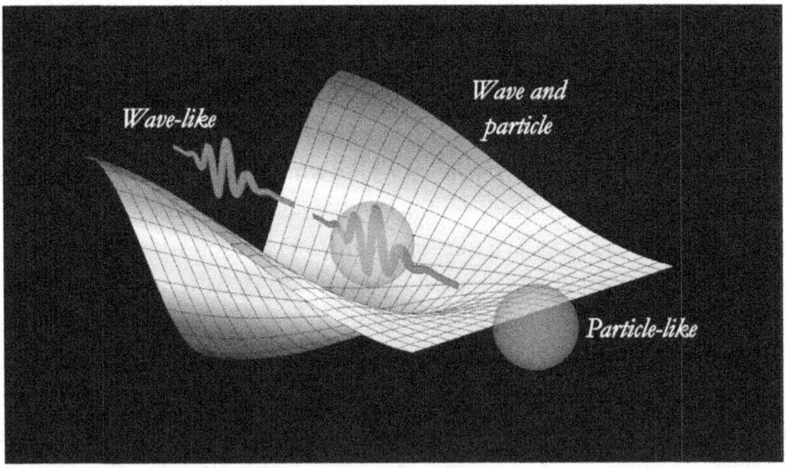

Photo Credit: S. Tanzilli, CNRS

Source: https://www.space.com/18346-light-wave-particle-duality-experiment.html

❄ ❄ ❄

5

Theory of Everything

PYTHAGOREAN THEOREM

The Pythagoras theorem is fundamental to physics as well as to mathematics. Although, it is generally credited to Pythagoras of Samos, Babylonian, Indian, and Chinese mathematicians knew some versions of it, many centuries earlier. Since then, so many mathematicians have offered new proofs of the theorem that it has been officially declared the world's "most proved theorem", in the Guinness World Book of Records.

Photo Credit: JAKNAPP 26

Source: https://commons.wikimedia.org/wiki/File:Pythagoras_knapp.jpg

The Pythagorean theorem first started in ancient Babylon and Egypt where the idea of the Pythagorean theorem had first been thought of. After the idea was proposed, the great mathematician

Pythagoras had taken this idea and made the mathematical base for this theory, and publicly stated the theory.

Pythagoras, born in 570 BC, formulated different theories and principles which were influenced by Aristotle and Plato, contributing to the revolution and development of mathematics, and soon published and proposed his Pythagorean theorem with the famous equation $a^2 + b^2 = c^2$ (where a, and b are the sides and c is the hypotenuse of the a right-angled triangle). Pythagoras is credited with many theories and discoveries, including the Pythagorean theorem, discovering the sphericity of the earth, how planets do not appear in morning daylight, and as well as the discovery of Venus. Not only did he make discoveries in science, but he also developed religious and cultural perspectives. He also developed music theory, and many of his students who were disciples of the Pythagoras, were called The Pythagoreans. His disciples also contributed to the works of music theory. He had discovered that musical notes could be translated into mathematical language and equations.

Since most of this universe is based on rectangles and right triangles, the Pythagorean Theorem has a very special and important place in geometry. Many relevant contemporary fields use applications of the Pythagorean Theorem. Architects use it to calculate the heights of buildings and the lengths of walls. Athletes such as basketball and baseball players even use the Pythagorean Theorem when they are calculating distance, to determine how fast they can run or where a ball needs to be thrown. Engineers and astronomers, use it to calculate the paths of spacecraft, including rockets and satellites.

Here, in this chapter, the focus is on how the Pythagorean Theorem has contributed to astronomy. Pythagoras taught that the earth and all heavenly objects were spheres. He reasoned this by looking at the moon. The light shining on the moon comes from reflected sunlight, with the moon merely acts as a projection screen. He observed that there is a curve to the zone between light and dark (the terminator) when it approached the edge of the moon. Only a spherical object would produce this curvature. If the moon were a sphere, must not all heavenly objects (including the earth) be spheres?

Pythagoras' Theorem also helps us calculate distances between the planets and stars. Even though we do not use the direct version, we more than likely used the modified version of the theorem, known as the Law of Cosines.

GRAVITY

The most important force of nature that has uncovered most of how things work on earth, is gravity. We rely so much on the force of gravity that the urge to discover more about it has become immense. But how did the idea of gravity come upon? Well, the legend says that the Great Isaac Newton discovered gravity when he saw an apple falling from a tree and inquired how it happened, why it happened, what caused it to fall, and so many more questions that we now have the answers to. Newton had been thinking about the forces of nature at the time and he directly linked his thoughts to the force that pulled the apple down, and eventually created a mathematical base, theories, papers, equations and so many more to prove this phenomenon. He later on concluded that an unknown force was acting upon the apple, causing it to fall or else it would not start moving and would stay at rest. After he proved this, he went on to create his three laws of motion that corresponded to his theories on gravity. His first law called the law of inertia, that states that a body is at rest unless acted upon by a force at a constant speed and when not in motion, it is still at rest. His second law is **"F = ma"** where F stands for force, m is the mass and a is acceleration. This describes the mathematical base of what happens when a force acts on a body. His last law of motion states that 'Every action has an equal or opposite reaction' which means that whenever the force is acting upon the object, it will cause either a opposite reaction of what it gave or an equal one. After he had published his new theories, he created Newton's law of universal gravitation which states that 'Any particle of matter in the universe attracts any other particle, with a force varying directly as the product of their mass and inversely as the square of the distance between them.' This statement is an advanced explanation of what he had first started with and gives the framework of the gravitational force. It means every particle is acted upon by a force due to its surrounding particles.

An AI-generated image of Leaning Tower of Pisa and an apple.

Source: https://gencraft.com/index.html

Way back before Newton was born, an experiment was done by Galileo called **"The Leaning Tower of Pisa Experiment"**, where he held two objects of unequal mass and dropped them to see which one would land first. Surprisingly, both had landed at the same time; this experiment had influenced Newton's work on gravity which is how he had concluded the Law of Gravitation as well. That was essentially how Newton uncovered the truth about gravity. However, ideas about gravity existed before Newton introduced his laws. Did all these new findings conclude that gravity is a force of nature that pulls objects toward the Earth? Well, before Newton many physicists had called gravity a weight or heaviness, and they used to also understand how the rotations of planets worked or how the moon orbited the earth and so on. Newton was indeed the first to use the word "gravity" to describe the force of nature, but the concept of gravity was approached quite differently in the Roman world. In that era, Aristotle taught that 'there is no effect or motion without a cause'. But his meaning of 'cause' was how heavy the bodies weighed, that corresponded to the downward movement of the motion or force. Another theory was formed within the Indian subcontinent, where they would believe people and objects all stood upright, but heavy objects and masses all fell to the ground towards the earth. They also believed that the elements on earth like water, soil, air all followed the same law which left them to wonder even more about how this worked. Many other regions, including the Islamic world, expressed curiosity; however, it was during the European Renaissance that Newton emerged to elucidate the mathematics and science underlying their questions and theories. We must mention the great pioneers of gravitational theory, starting with Aristotle, Galileo, Newton, and the great Albert Einstein, all of whom have significantly influenced the works and ideas of gravity. But why is understanding gravity in-depth important for us and humanity? There could be other factors we could focus on instead of gravity since Newton, Einstein other great scientists have already explained most of what gravity is, and how it behaves. Well, gravity is one of the main factors on Earth that makes it habitable; without it, we would not be able to survive. We have witnessed numerous mathematical proofs and theories explaining how gravity works. However, delving deeper into the understanding of gravity can expand our knowledge of space travel, aid in understanding other planets, and potentially unlock new insights into the universe that were previously unimaginable. Gravity may be an essential tool for everyday living, but it can be way more important in discovering and learning new things about how gravity behaves or interacts with other forces and masses; gravity describes how all matter interacts and behaves by giving the force of gravity and maybe even beyond. Gravity influences our space travel by guiding the paths taken by satellites or rockets, debris, or anything traveling through space, so it is important to know which path one of our creations may take to reach another planet or a distance that may be far beyond our capacity. By understanding gravity, we can use that knowledge and calculate how we may send a satellite to reach a certain destination, without any disturbances or interruptions going within space; many discoveries and inventions could happen within this. Recent theories have all been linked or related to gravity which gives us more to understand how gravity could be much more important to us, than we think. Some time ago, gravitational waves were discovered and observed by the Advanced Laser Interferometer Gravitational-wave

Observatory (LIGO). They detected the first gravitational wave, generated by the merger of two supermassive black holes. This event caused a ripple in space that travelled through both space and time, eventually passing by Earth in a matter of nanosecond. LIGO's advanced scientific technology played a crucial role in capturing and analysing this extraordinary phenomenon.

Now that we have finally understood the fundamentals of gravity, we shall discuss the discoveries that have taken place to enhance our knowledge about gravity. A discovery about gravity that has recently come out is how the level of gravity in a certain place on the Earth can help predict an earthquake or other natural disasters. Another one is how gravity can help us understand better about the accelerated expansion of our universe. We are yet to find out the mystery behind this expansion. Is it due to the presence of a mysterious exotic matter, normally referred to as dark energy, or does indicate the breakdown of Einstein's gravity at a large scale? Among the two, the modification of Einstein's gravity is more reasonable, as with the same theoretical framework, one can also describe the early epoch of cosmic acceleration, *aka* Inflation. Modified theories of gravity provide a wide variety of possible models for a correct description of the universe. Gravity plays a crucial part in all our lives including the universe we live in.

Photo Credit: Alfred Pasieka/Science Source

Source : https://www.quantamagazine.org/gravitational-waves-should-permanently-distort-space-time-20211208/

GENERAL THEORY OF RELATIVITY

The general theory of relativity is the theory of gravity developed during 1907 to 1915, by Albert Einstein. The general theory of relativity states that 'The observed gravitational attraction between masses results from the warping of space and time by those masses.' It essentially means that any body in our universe is surrounded by a fabric that we call space-time. The celestial bodies such as our sun or black holes, planet and stars all sit on top of this fabric, and depending on how heavy the collective mass, the more it warps the fabric around it. Very dense bodies, such as neutron stars, warp the space-time fabric immensely, creating a deep hollow. With even a slight increase in mass, the space-time might undergo further distortion or potentially break. The theory of General Relativity that Einstein published in 1915 extended the theory of Special Relativity that he had formulated a decade earlier. Special relativity argued that space and time are intricately connected, but it did not acknowledge the existence of gravity. Several physicists, including Einstein, searched for a theory that would reconcile Newton's law of gravity and special relativity. Over the years, Einstein refined and scrutinised his mathematical expressions until 1912, when his friend from ETH Zurich, Marcel Grossman, introduced him to a new mathematical tool called tensor analysis. This new approach allowed him to express his laws of physics and formulate his new theory of General Relativity, which was ultimately published in 1915. This groundbreaking theory revolutionized the entire field of science and reshaped our understanding of the universe.

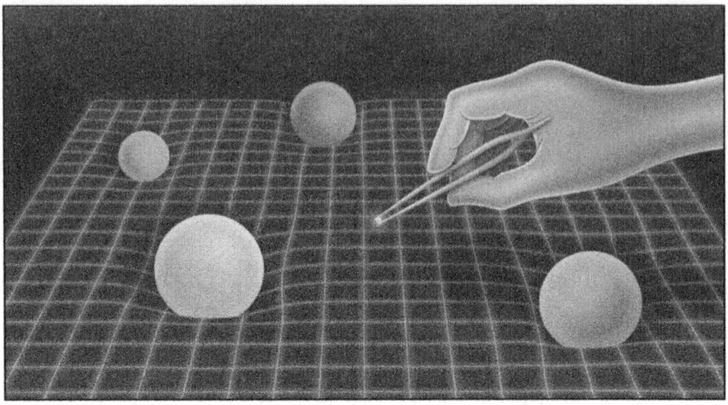

Photo Credit: Kristina Armitage/Quanta Magazine

Source: https://www.quantamagazine.org/a-century-later-new-math-smooths-out-general-relativity-20231130/

But why is the general theory of relativity important for space exploration and today's physics? Well, the General Theory of Relativity has proven invaluable in both practical applications and research. It enables us to make predictions on the expansion of the universe, understanding how black holes, gravitational waves, and many more unknown objects and masses could appear on the space-time fabric, distorting time itself. The general theory of relativity has produced a lot of solutions and technologies we use every single day. Einstein explained why clocks on satellites are ahead of those on Earth by 38,000 nanoseconds. Our GPS navigators which take the distance of one point of the Earth to different satellites account for this. It doesn't seem like much, but if these nanoseconds weren't accounted for, GPS would not be accurate and would be off by a couple of kilometers or more.

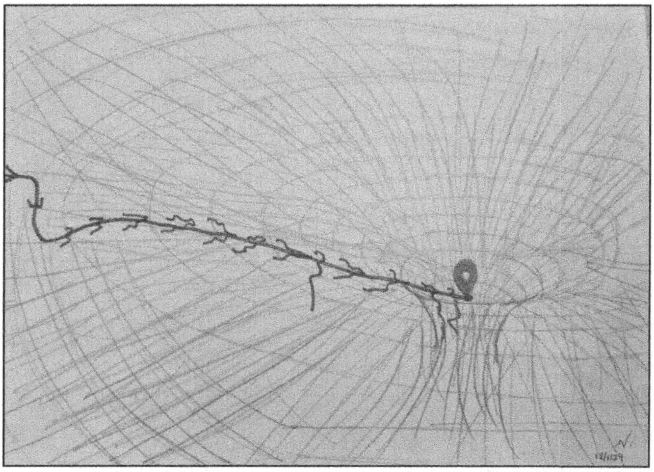

Photo Credit: Oindrila Majumdar

MAXWELL'S EQUATION

Though Maxwell did not discover electricity and magnetism, he did put in place a mathematical formulation of electricity and magnetism that was built upon the earlier work of Benjamin Franklin, André-Marie Ampère, and Michael Faraday.

Photot Credit: 3Blue1Brown

Source: https://www.youtube.com/watch?app=desktop&v=rB83DpBJQsE (minute: 1:33)

To understand Maxwell's equations and their far-reaching consequences on us and our universe, which we are still exploring, we need to go back to 1865, when electricity and magnetism were two separate branches of physics, governed by separate laws of physics. The electric and magnetic forces were two unrelated strange invisible forces. Andre-Marie Ampere and Michael Faraday were the first to link electricity, magnetism, and light. In 1845, through a complex experiment, he demonstrated that a strong magnetic field could change the polarisation of a light ray, which suggests that magnetic force and light were related. By the time Maxwell joined the research, Ampere, Faraday, and their predecessors had formulated various laws that hinted at the links between electricity and magnetism. But nothing connected these ideas. It took 10 years for Maxwell to set about mathematically describing Faraday's lines of force, to account for all the electric and magnetic effects that were observed by then. He built a theory of electromagnetic fields. His complete set of equations governing electricity and magnetism expressed in terms of fields was summarised as Maxwell's equations in 'A dynamical theory of Electromagnetic Field', which was published in 1865. Institutively, the equations can be summarised as something like:

Electric and magnetic fields make electric charges move. Electric charges cause electric fields, but there are no magnetic charges. Changes in magnetic fields cause electric fields, and vice versa.

Photo Credit: Institute of Physics, UK

Source: https://www.iop.org/explore-physics/big-ideas-physics/maxwells-equations#gref

From his electromagnetic wave theory, Maxwell calculated the speed of the electromagnetic wave in the air, which he saw was near the speed of light. He, therefore, concluded that light itself is an electromagnetic wave. This was a revolution in the scientific understanding of light and, consequently, the greatest advance since Newton's Principia. Maxwell applied the application of mathematics to physical laws. Without sophisticated telescopes, and by simply using mathematics he proved that Saturn's rings were made up of millions of loose pieces, as that is the only way the rings could be stable. This was observed 120 years later, in 1980 when the Voyager spacecraft captured a detailed picture of Saturn's rings, comprising of spinning particles of dust, stones, and huge rocks. The electromagnetic wave had a huge impact on technological advancement, as in 1888, Hertz discovered other types of electromagnetic waves which had applications in radio communications, television, radar, and navigational aids, and more recently in Wi-Fi signals and smart phones, that allow communication in ways not dreamt of, a generation ago.

Maxwell's equation had a huge role to play in astrophysics, both in theoretical advancement, as well as in experimental observations. Theoretically, we know, one of the two fundamental axioms from which the special relativity theory derives its mathematical hypotheses, which is also experimentally confirmed, and it is that the speed of light, or more formally, the speed of electromagnetic waves is constant and frame-invariant. This amounts to say that in all frames of reference the speed of light is constant, and with the same value, everywhere. This axiom was proposed because it is the only possible interpretation of Maxwell's equations, which assumes that Maxwell's equations are a complete description of classical electromagnetism. If Maxwell's equations were false, then there would be errors in the theory of special relativity too, and astrophysics would have been completely different from what it is today.

In astrophysical observations, the entire electromagnetic spectrum is used to observe a variety of things in the universe. Radio waves and microwaves, which are the longest wavelengths, are used to look inside dense interstellar clouds and track the motion of cold, dark gas. Radio telescopes are used to map the structure of our galaxy. Also, microwave telescopes are sensitive to the remnant glow of the Big Bang. Infrared telescopes excel at finding cool, dim stars, slicing through interstellar dust bands. Additionally, they measure temperatures and detect particles of planets and stars. The wavelengths of infrared light are long enough to pierce through clouds that would otherwise block our view. By using large infrared telescopes, astronomers look through the dust and warm gas of our galaxy into the Milky Way's core.

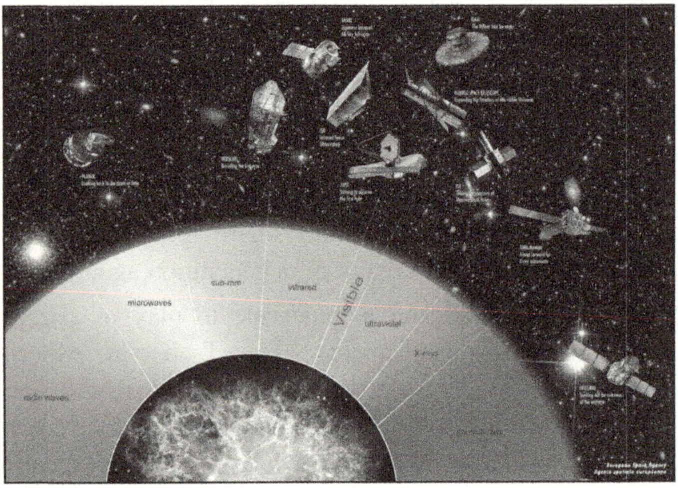

Photo Credit: European Space Agency (ESA)

Source: https://www.popularmechanics.com/science/a41473994/how-light-travels/

https://sci.esa.int/web/education/-/49055-esa-s-fleet-across-the-spectrum

The majority of stars emit the bulk of their electromagnetic energy, mostly as visible light, a tiny portion of the electromagnetic spectrum that our eyes can see. As the wavelength is related to energy, the colour of a star tells us how hot it is: red stars are cold, and blue stars are hot. The coldest stars do not emit visible light at all; they can only be seen with infrared telescopes.

The most energetic portion of the universe is the region of the formation of new stars. To identify these regions astronomers use short-wavelength ultraviolet, or UV, light. When viewing distant galaxies with UV telescopes, most of the low-energy stars and gas disappear, and all the stellar nurseries pop into view.

X-rays and gamma rays have the highest energies in the electromagnetic spectrum. X-rays come from the hottest gas that contains atoms, superheated spiraling particles that come from black holes or exotic neutron stars. On the other hand, gamma rays have the shortest wavelength, which helps unveil the most violent events, such as cosmic radioactive decay, supernova explosions, and even the destruction of antimatter. Gamma-ray bursts, brief flickers of gamma-ray light, are the most energetic singular events in the universe. They are observed emanating from active galaxies, pulsars, pulsar wind nebulae, and during the explosive formation of a black hole when a star undergoes a supernova.

Therefore, we see how astronomers extract information about the universe from the knowledge of electromagnetic waves and Maxwell's equations. Telescopes that we send into space like the Hubble telescope or the most recent James Webb Space Telescope (JWST), are constructed considering these equations. For example in JWST, why is Webb's mirror gold coated? This is because the unique atomic structure of gold reflects infrared light very well. Light being an electromagnetic wave, induces oscillations in the electrons near the surface of the gold, when it hits it. The surface electrons move such as to make the electric field zero. The infrared light is re-emitted or reflected away in an opposite direction, as it makes the net electric field of the moving electrons and the electric field of the light, adding up to zero, in the gold metal. This can be directly derived from Maxwell's equations. The first profound unification of nature's forces by Maxwell had huge contribution to mathematics, astronomy, and engineering, so he is rightly acclaimed as the father of modern physics. Maxwell's equations prompted Einstein's research in relativity. Einstein is quoted as saying:

"One scientific epoch ended and another began with James Clerk Maxwell".

"The Special theory of relativity owes its origins to Maxwell's equations of the electromagnetic field".

Photo Credit: Peter Greenwood for Quanta Magazine

Source: https://www.quantamagazine.org/how-to-think-about-relativitys-concept-of-space-time-20221114/

PARTICLE PHYSICS

Atoms, particles, and microorganisms form life and biological structures around us. Biologically, they help us discern the inner microscopic life that goes way deeper than what even a microscope can detect. This inner microscopic life exists way within the biological structure where we see how life is formed and how it functions inside. Many models of atomic behaviour and particles help us see how these micro lives behave and interact with each other. But particles, atoms, and molecules don't apply only to biology or chemistry, they also apply to astronomy, viz. how they may interact and behave in the mysterious and wavy space-time fabric. Particles in motion have always amazed scientists and how they interact with each other in different situations, such as how particles are tightly packed or carelessly move around, in the three states of matter.

Photo Credit: Photo courtesy of Giulio Biroli; alloy image from Sugimoto et al., 2007

Source: https://www.quantamagazine.org/solid-or-liquid-physicists-redefine-states-of-matter-20130403/

The origin of the universe was originally from particles. Many theories were based on particles like string theory, wave-particle duality, and many more, so it's no surprise that particles have always perplexed us, and we haven't yet uncovered the whole idea of particles. Scientists and physicists have done experiments and used particle accelerators to discover new unborn particles, and we will understand how they work after a while. The most fundamental way to describe our universe and what it's made of is the concept of particle physics. It describes the tiniest and most important factors in our universe to learn how different forces, particles, and masses behave with each other based on what it is made of. Quantum mechanics also play a big role in looking into the macroscopic world of the universe and beyond, but that's a whole other concept in astrophysics for another time. The first discovery that particles exist, and it makes up most of our world is in the 1930s when a standard model had been created to show all the fundamental particles or 'Elementary particles', which contained the Electron, Proton,

Neutron (inside the nucleus), Neutrino (now anti-neutrino) and a photon (the quantum side within the electromagnetic field). These are basic particles which we know by now make up our basic knowledge of what matter is made of. In 1932, physicist Carl Anderson discovered the positive electron (positron), which also largely contributed to knowing more about particles and matter. Later, in 1932, James Chadwick discovered the neutron which had made up most of an atom. Over the years, many other developments had happened in discovering different parts of particle physics, in slowly furthering our understanding of the atom. In 1935 Hideki Yukawa, a Japanese physicist, had proposed Yukawa's meson hypothesis which was about the nuclear force in the particle interaction and exchange within masses (mesons). Much later, in 1950, history was revolutionized when the Quantum Theory of Electromagnetism or Quantum Electrodynamics (QED), the theory in which 'the interactions of charged particles in an electromagnetic field' was introduced by Julian Schwinger, Sin-itro Tomonaga, and Richard Feynman. This had brought the idea of quantum mechanics as well. 5 years later the idea of anti-matter and the anti-proton was introduced by physicist Owen Chamberlain and Emilio Segre. After that many other developments of particle physics and more had been initiated and published until evidence of the revolutionary Higgs Boson particle was discovered from organizations like LEP and CERN through a particle accelerator in 2000, but it wasn't yet confirmed back then. Now CERN used the Large Hadron Collider (LHC), the world's largest and most powerful particle accelerator, which is used for allowing physicists to test theories involved in particle theories. Now, they are in search of the dark matter particle, since dark matter is an invisible force that we know exists because of its impact on different parts of the universe. But we cannot understand what kind of particles it is made of since we cannot see it. However, we do know what it isn't. We cannot quite understand what type of particles make up the dark matter, because it doesn't interact with normal matter, but CERN is in the process of trying to find a suitable particle that fits into the description of Dark matter. Particle physics is the building blocks of our universe and the speed at which new technologies are emerging, we can uncover more about the deepest and darkest parts of the universe.

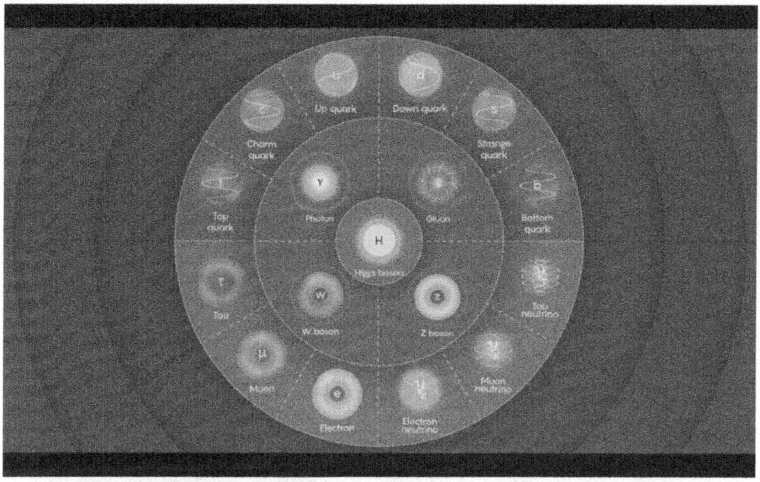

Photo Credit: Quanta Magazine

Source: https://www.quantamagazine.org/solid-or-liquid-physicists-redefine-states-of-matter-20130403/ (minute: 13:04)

DARK MATTER

Imagine taking a hike up on the mountains in the early morning and finally reaching a point where everything is covered in fog and mist, and you start to lose your path. You know there are rocks up ahead, but you do not know what type of rocks, they could be stable, or unstable, and that's exactly what approaching dark matter is like. You know there is a force of dark matter, but you cannot see it, or you do not know what it's made up of, but there has been so much evidence that it exists and the only thing we know about dark matter is what it isn't. The world is currently trying to understand what dark matter may be made of, because we do not know what the composition of dark matter is and what it consists of because it does not interact with normal matter that is familiar to us. There are many possibilities on what kind of particles dark matter may be made up of, such as a neutrino, a particle with a negative charge which, however, is unstable, which means it moves around, and dark matter particles are required to be stable and unable to interact with each other. So far, recent developments to find new hypothetical particles that could match the requirement for dark matter, like neutrinos almost do, are Weakly Interacting Massive Particles (WIMPs), which mostly fit with dark matter particles and super-symmetric particles which also makes them good candidates.

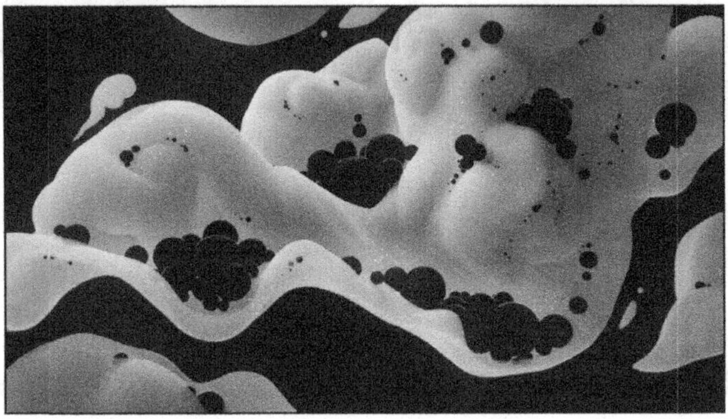

Photo Credit: Olena Shmahalo/Quanta Magazine

Source: https://www.quantamagazine.org/black-holes-from-the-big-bang-could-be-the-dark-matter-20200923/

Recently, many headlines have come from the huge Large Hadron Collider (LHC) at CERN in Geneva, Switzerland which collides and accelerates particles in the hope to discover or create a new particle, as they had done for the Higgs Boson. The LHC has recently come across three new exotic particles. This kind of state-of-the-art equipment could be used to create a dark matter particle. Even if we find a suitable particle or make one, we wouldn't be able to create a real particle out of a particle accelerator, because a dark matter particle should be extremely light and dense so it would immediately escape. However, all hope is not lost. Luckily light energy would be left which we could observe and analyze and somehow find a suitable candidate that suits the description of dark matter. Hopefully, we will soon be able to find such a particle to help us understand the depths of dark matter. Since we know that dark matter makes up almost 27% of our universe, by determining the amount of dark matter and where it is, we can know if the universe will continue to expand or come at a point where it stops and collapses altogether. But the real question is how do we know dark matter even exists? Well, there are various ways such as gravitational lensing, which is used to see distant galaxies through a cluster of galaxies. In dark matter, gravitational lensing is used to observe massive galaxy clusters and we can see how dark matter bends and distorts the light of more distant galaxies behind the massive galaxy cluster. Also, astronomers can map out certain areas where dark matter may exist, by looking at the distortions of light, as well. We also know that dark matter causes galaxies to spin, one incident that astronomers and physicists have seen is that galaxies gradually spin faster at the point where stars should be ejected. However, they could see stars being upheld by some kind of strange force and eventually they said dark matter was the reason. But how can we detect it and understand it? Understanding gravity better could enhance the knowledge of dark matter as, on some occasions, dark matter is claimed to be acting like anti-gravity. Another way to detect dark matter is knowing where in the universe more galaxies are being formed, as dark matter may also be part of the process, as well. Quantum mechanics and the General Theory of Relativity can be used to find a new particle for dark matter, since quantum mechanics has a lot to do with the micro-world and the General Theory of Relativity is used to find evidence on dark matter. Einstein himself talked about dark matter: "Most of the matter in the universe is dark matter", which shows that even the General Theory of Relativity can be used to understand dark matter and what it's made of. Finding dark matter will be possible with the new advanced technology and maybe in the future we would find a new particle that fits exactly into the dark matter description and maybe it can solve the problems of our never-ending questions about dark matter and the other unknown objects in our universe!

BIG BANG

Have you ever wondered what it was like before our universe was born? How did it come to be what it is now? Are there other universes like ours out there? There is endless curiosity about how our universe began, but we've only been able to come up with one theory that most people have settled with, viz. The Big Bang. Many people like to stay with the idea of believing that two particles clashed together and suddenly boom, our universe came to life. However, this is just a theory and there could be many such possible explanations of our universe and its origin, but we are extremely close to finding the past of our universe with the advanced technologies we have today. But how did we come to the idea of the Big Bang and what is it? Well, let's start with the definition of the Big Bang. The big bang theory is the theory of how our universe first started and expanded from a state of density and temperature. 13.8 billion years ago there was a ton of energy jammed into a very tiny and dense point which eventually exploded with an incredible force and created matter, which eventually made billions of galaxies and stars, while vastly expanding into deep space and eventually creating our milky way galaxy, the stars in it, our solar system and then our earth.

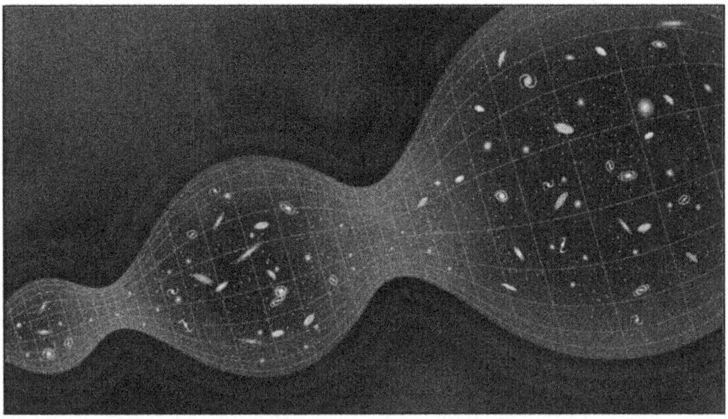

Photo Credit: Illustration: Samuel Velasco/Quanta Magazine

Source: https://www.wired.com/story/what-if-the-big-bang-was-actually-a-big-bounce/

Then there are other questions: Are there more universes? Did they also originate through the big bang, like our universe? Are there multiverses? This all brings us back to our first theory of the big bang. Maybe in different multiverses, the laws may be different, for instance, light may behave in a different way or gravity works the other way around, maybe black holes could turn into stars and there are so many possibilities that the concept of the Big Bang opens up.

Apart from our understanding of the big bang, now there other theories that suggest how the universe first started. According to a new theory of the universe, time and space may not have started with the Big Bang. but rather may have always existed in an endless loop of expansion and rebirth. Many such theories exist and we are still trying to find an answer. Back in the day, many people believed that it was a god who had created existence, but now that other explanations, exist we wonder more about the origin of the universe. Currently, new telescopes have made it possible to investigate our universe's past, to see how galaxies formed when the universe was still young, how different stars died and formed into black holes which we now know today and many more. If it is possible to look beyond 13.6 billion light years into the past, we would be able to see our early universe forming. The big bang may or may not be true and maybe a close estimate of how the universe formed, and no later in a few years we'll be able to unlock the universe's darkest secrets.

TIME

Time is a concept that has changed over centuries, many people believe that time is just a 'subjective illusion of human consciousness, that the past, present and future are equally real, and time is senseless', say B-theorists; our real understanding of time comes from the great Albert Einstein. Before him, Newton had also worked on time and said time was a constant meaning, wherever you go time will be the same- a notion which has now been proven wrong. Einstein's theory of time has put forward that time is relative, depending on the gravity in some places time may slow down or go faster than others. 100 years later, Einstein is still correct. Many people have tried to test his theory through the Hafele-Keating experiment, which places a cesium-beam atomic clock on a plane and another clock on the ground and to see if time would remain the same or if it would show different times. It indeed showed different times; whereas the plane's time was way faster since at the plane's altitude and so sped up, since there is very weak gravity. This was the gravitational time dilatation that Einstein predicted in his theory of General Relativity. It is proven that time slows down near a black hole because of its massive gravitational field; it is said that 1 second near a black hole is equivalent to 100000 years on our earth, but it still hasn't been proven yet. Time is a very interesting topic and can make your brain twist and turn with all the complex theories and questions we ask about time.

Photo Credit: Janelle Barone/New Scientist

Source: https://www.newscientist.com/article/mg25534011-500-why-rethinking-time-in-quantum-mechanics-could-help-us-unite-physics/

Back in ancient Egypt, they first created the idea of time by creating a sundial that used the junctures of sunrise and sunset, and put the period of the sun rising and setting into twelve parts, showing the hours between these two. Many theoretical physicists and scientists say that time may be an illusion and according to theoretical physicist Carlo Rovelli, time is an illusion, because the flow of time doesn't exactly correspond to the physical realm, making Einstein say that time could be the fourth dimension, and nothing definite could be said about past, present and future. There are many theories as to whether we can stop time (Yes it is theoretically possible, if you travel at the speed of light). Is time infinite? is another question among a host of others, to be answered. There is a theory Einstein formulated called space-time or space and time, which is where all bodies in space have a certain mass and they are all sitting on top of a fabric (which represents the space and time), and the fabric warps and distorts around the body depending on how much mass it has. For a black hole, the fabric warps completely and eventually lead to singularity. Another kind of mass is a neutron star which is a very dense mass that warps space and time and touches the end of the distortion. If a little denser mass is added to the star, it may break into space-time and create a wormhole that can bring us to a different dimension or maybe time travel! Time can be an extremely complex concept that will be studied for years to come because of its perplexing properties and questions.

THE MULTIVERSE

Photo Credit: Peter Greenwood for Quanta Magazine

Source: https://www.quantamagazine.org/are-there-reasons-to-believe-in-a-multiverse-20230517/

The multiverse is a theory that has been going around for a while, about how multiple universes exist other than ours and there could be millions more or billions. Understanding the multiverse will be difficult since we don't even understand or know 97% of our universe, but predicting from what we know we have come up with questions and theories that may start a journey to understanding what may be beyond our universe. The idea of the multiverse appears in a few branches of physics (and philosophy), but inflation theory provides the best-known example. According to inflation theory, a hypothetical occurrence took place when our universe was young - less than a second old. According to NASA, the cosmos experienced a brief period of fast expansion before "inflating" to become the universe it is today. But how do we know the multiverse exists? Have we found any solid information or data that proves the multiverse is real? Well, we do not have any evidence that proves the multiverse may exist, but the existence of the multiverse is just a theory that our universe is not alone, and it is a bubble among many other bubbles that continue to expand and start continuously, even if we may not know if another universe exists, we could predict and try to theorize how other universes may function. There is no way that all universes will be the same, maybe each universe started in different ways unlike ours with the big bang, other universes could have a different law of gravity or physics, and maybe light only goes in a straight line and doesn't bend, or maybe that universe only contains stars or galaxies and not the other element that we have in our universe, so all we can do now is try to come up with theories on how different universes behave. 'But right now, the idea of the multiverse is purely theoretical and, in some cases, philosophical' according to National Geographic, but maybe in the future we may be able to understand more about if there is enough evidence about the multiverse. For now, scientists and physicists have come up with three possible types of the universe. Type one is the flat universe where there is no dimension and zero curvature; the second type is a spherical or curved universe something like ours (positive curvature); and finally, the last type which is a hyperbolic or open universe which has negative curvature. There could be many more types of universes, but for now we have come up with only three which gives us a good idea of how other universes may behave and interact. The multiverse will be a theory explored in every way, but now we still have 97% of the universe to explore, once we get closer to knowing more about the universe, we can get more about different universes out there like a huge leap into looking through the past of the universe when the universe first started, which can determine if the universe started through the big bang. There is much research being done and before we know it, we will know more than 3% of the universe and beyond hopefully in the next few decades.

MODIFIED NEWTONIAN DYNAMICS (MOND)

Newton's law of gravitation was about how gravity behaved and how its particles or gravity's force acts with other forces, and how they affect us. But there is also a modified version of Newton's universal law of gravitation called Modified Newtonian Dynamics (MOND), but it is a hypothesis that was proposed as the modified version of his idea of gravity, which is specifically for observing different properties of galaxies. It is also known for being an alternative for dark matter as well, since galaxies and dark matter have a connection through how galaxies can spin (the reason being dark matter). Also, looking into new modified gravity theories we can find solutions to the existence of dark matter and dark energy.

Photo Credit: NASA/JPL-Caltech/ESA/CXC/STScI

Source: https://www.quantamagazine.org/modified-gravity-theory-passes-a-critical-test-20200728/

The idea of MOND was created in 1982 and first published in 1983 by Israeli physicist Mordehai Milgrom who had said that the original motivation was to explain why the velocities of stars in galaxies were observed to be larger than expected based on Newtonian mechanics. Gravitational lensing is another aspect that MOND may deal with, since gravitational lensing is a way to detect the existence of dark matter. MOND is another alternative model for gravity which many scientists and physicists use to try to find out how gravity can be modified or written and understood differently.

This helps because our previous theory, the multiverse, says that in every multiverse the laws change and everything behaves and acts differently, so studying about the modified version of Newtonian dynamics or gravity it helps us understand how gravity could act in other universes, which bring us a step closer to knowing if there is a multiverse out there. MOND includes a lot of mathematical bases and formulas which cannot yet be said in layman's terms, but we can still get an idea of what MOND is trying to do.

In the name Modified Newtonian Dynamics, we know the main topic is Newtonian Dynamics. Newton's laws of dynamics show us the dynamics or mechanism of particles and smaller bodies according to Newton's laws of motion, but MOND helps us find more alternative ways that particles and smaller bodies can be modified eventually. MOND will be better understood when we get more research and theories regarding the subject. But for now, all we know is we are trying to find different models to represent gravity which will probably help us in the future and ahead.

COSMIC MICROWAVE BACKGROUND (CMB)

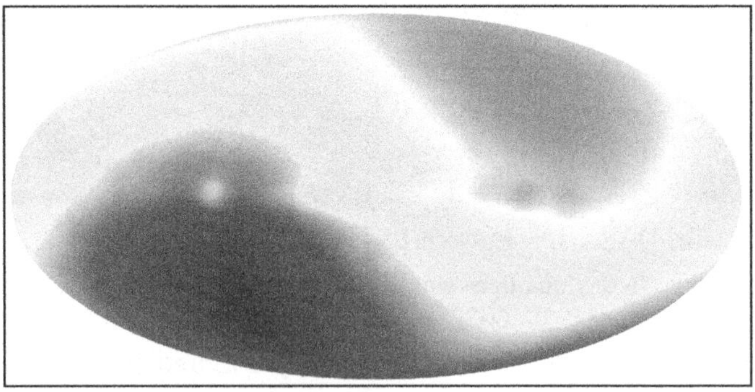

Photo Credit: NASA

Source: https://www.quantamagazine.org/giant-arc-of-galaxies-puts-basic-cosmology-under-scrutiny-20211213/

The Cosmic Microwave Background (CMB) is about how further back we can look in time and explore using light, and we know the James Webb Space Telescope is already doing this by looking at stars and galaxies when the universe was just born, which is possible by how long the light of the certain body's light takes to reach us; maybe we may even be able to see the beginning of the universe, but that for another story. The CMB also helps us form a dark matter map, and it can detect and see in which areas our universe may contain dark matter; through it we can also see the unlimited amount of stars and galaxies being formed over time; thermal lensing is used to see which areas of the universe may be hot or cold and we can also see the variations of temperature as well, last. We can also see our expansion of the universe, but the CMB mostly focuses on the exploration through light and how our universe first started. The CMB was created by the leftover radiation from the big bang, when the universe had just started and was created. It also shows evidence of how quickly the universe expanded and evolved from the big bang. The CMB was discovered and created roughly 400,000 years after the Big Bang, 13.8 billion years after the universe's creation. That's because, according to NASA, the temperature of the early universe, when it was just one hundred millionth the size it is now, had extremely high temperatures, when it first began. Any atoms that were there at the moment were soon divided into tiny particles (protons and electrons), which is where we get the known particles that make up an atom. The electrons were spread across photons from the CMB, which are particles that represent the quantum states of light or other types of radiation. These particles soon came into the composition of atoms which almost make up everything that exists in the universe. Then, photons travelled across the early universe, according to a NASA report. The universe was cool enough 380,000 years after the Big Bang for hydrogen to exist, and the universe started expanding rapidly and soon created many more gases and particles which now made our universe today, which continues expanding and growing phenomenally to this day. However, the CMB was stumbled upon initially, by Arno Penzias and Robert Wilson, two researchers at Bell Telephone Laboratories, who were developing a radio receiver in 1965 and were confused by the noise it was picking up. They soon discovered that the noise was coming from the entire sky. At the same time, a Princeton University team under the direction of Robert

Dicke was looking for the CMB. After learning about the Bell experiment, Dicke's team recognized the CMB had been discovered. In 1965, both teams had successfully published publications in the Astrophysical Journal, with Penzias and Wilson discussing what they had seen and Dicke's team described the cosmic significance on the importance of this big discovery of the Cosmic Microwave Background (Afterward, Penzias and Wilson shared the 1978 Nobel Prize for Physics). The CMB is important for us to understand and learn how the early universe or when our universe began was formed. The CMB model is also the earliest picture of the universe we might have where light had helped us take the legendary picture.

HIGHER DIMENSIONS

Photo Credit: Maggie Chiang for Quanta Magazine

Source: https://www.quantamagazine.org/a-mathematicians-guided-tour-through-high-dimensions-20210913/

The normal dimensions for space for us as well, are only three, which are length, width, and height, but for space, it would be 4, the fourth one being time. The highest dimension of the universe can have been 11 dimensions, higher than 11 would collapse into an 11-dimensional universe, eventually. Physics researchers believe that the proposed theory of everything, known as string theory, will eventually explain everything. Under a single theoretical roof, where everything is vibrating threads, are all the forces, particles, constants, and everything. Theorists have been working on the idea since the 1960s, and one of the first things they noticed was that there needed to be more dimensions than the four, that we are accustomed to for the theory to make sense. According to string theory, the different particles (such as electrons, quarks, neutrinos, etc.) which are the main components of today's atoms and masses and the force-carriers of nature are manifestations of tiny loops of vibrating stringiness, which the theory claims are the fundamental object of reality (photons, gluons, gravitons, etc.). They accomplish this by vibrating, which is how, each string is so little that it only appears to us as a point-like particle, but it is capable of vibrating in a variety of modes, just like a guitar string can produce a variety of notes. Each vibration mode is believed to correspond to a certain type of particle. Thus, all strings vibrating in one direction appear to be electrons, all strings vibrating in another direction appear to be photons, and so on. We can't reasonably hope to directly explore these with our high-energy tests due to the very small size. And if those additional dimensions are concentric with one another, then every time you travel in four-dimensional space, you are navigating those extra dimensions, billions upon billions of times. And those are the dimensions in which the string theory's strings exist. With more mathematical understanding, it was discovered that the additional six spatial dimensions required by string theory must be contained in a specific collection of configurations known as Calabi-Yao manifolds, named for two eminent physicists. However, the string theory does not admit a single distinct manifold. Higher dimensions connect with the string theory which talks about how every particle has its stringiness which evolves a particle and through that the higher dimensions on the dimensions of the universe and the composition of particles as well. Knowing higher dimensions slowly brings us to how the universe was first created and how it is continuously expanding as well, maybe as it continues to inflate and expand, we could come up with a higher dimension than 11 which may even become a whole new universe as well. String theory also plays a big role in understanding higher dimensions as string theory is claimed to be the theory of everything.

STRING THEORY

String theory is claimed to be the king of all theories and it explains how new particles are made and born through vibrations of strings. There is a debate if either string theory should be called the theory of everything since everything is made up of particles and string theory says that the particles are found in vibrating strings. I agree with calling it the theory of everything, since it includes topics like higher dimensions and particle physics, but before jumping to conclusions, what is string theory in the first place? According to the theoretical physics concept of "string theory," reality is composed of tiny vibrating strings that are smaller than atoms, electrons, or quarks. By this idea, the strings' vibrations, twists, and folds result in effects in numerous, microscopic dimensions that people can interpret as anything from particle physics to more significant phenomena, like gravity. General relativity and quantum mechanics, the two theories that form the basis of practically all modern physics, have been brought together by string theory as a potential "theory of everything." Both general relativity and quantum physics are excellent at explaining the behaviour of extremely small objects, but they clash when trying to explain the behaviour of very massive objects in the universe. According to some researchers, string theory has the potential to explain the puzzles that exist between the two, solving one of the most difficult physics problems that have yet to be addressed. In Albert Einstein's general relativity theory, gravity is a force that distorts the space-time of heavy objects. Gravity is one of the four forces used by physicists to describe the natural world. However, gravity is so weak compared to the other forces (electromagnetism, the strong force, and the weak force) that it cannot be detected or observed at the level of even a particle. In contrast, its effects are only significant when seen in terms of planets, stars, and galaxies. Additionally, gravity does not appear to be a separate particle. Theorists can predict the shape of a gravity particle, but when they attempt to predict what happens when two such "gravitons" collide, they obtain an unlimited amount of energy compressed into a tiny space - a black hole. Every force has its particle which can also be explained by the standard model by looking at and understanding string theory. String theory consists of many topics which can help us understand more about string theory, like particle physics and the particle side of quantum mechanics, and how they interact in higher dimensions. Looking at how they behave we can see how the strings vibrate and how they can produce new particles or how they vibrate in different dimensions, such as the 4th dimension. String theory can be thought of in many ways which help us understand to go beyond particles and maybe even beyond vibrations of strings.

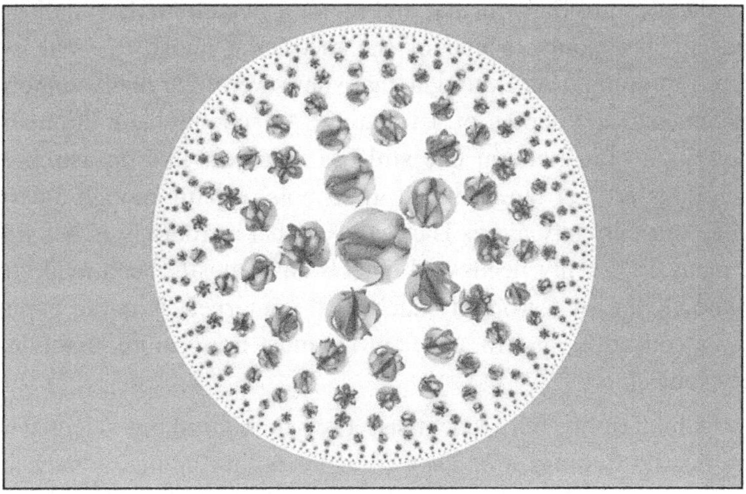

Photo Credit: Glen Faught for Quanta Magazine

Source: https://www.quantamagazine.org/string-theory-only-game-in-town-tests-20150218/

CONCLUSION

Back in 1920, physicist Georges Lemaître had created the first theory on the universe, the Big Bang theory. The Big Bang theory has helped us uncover many aspects of our knowledge of the universe and many other theories were born by renowned scientists which gave us more to think about, and then more theories get created based on another theory and so on. Theories can help us validate whether any other theory can be proven or not for any research. There is a plethora of theories that have been proven, but many are still not known due to the limit of advanced technology we currently have. The multiverse theory, Time travel theory, or maybe even certain paradoxes haven't been understood yet, but we can hypothetically imagine how the universe would work or how time would work if all these theories were to be proven. But coming up with new theories helps us prove other theories as well, and through experiments and research to discover our universe's true nature. Coming to this chapter, I have chosen a specific number of theories such as gravity, higher dimensions, string theory, etc., and most of these theories are fundamental and the first steps to understanding the basics of astronomy. Some of the theories I chose to write were purely my opinion as I thought that some were extremely important to understand the depths of astronomy. As schools have drained the curiosity and imagination one still has a subject like on astronomy, which does not get taught much in schools or paid enough attention to keep our curiosity alive. The reality is that astronomy is our future and many such daily needs and objects have mostly been made from the knowledge of astronomy such as the internet. A quote explains today's education system and how we must bring more creativity to every subject.

"Children are born scientists. They do everything scientists do. They test how things are. They measure falling bodies.... they learn the physics of the world around them. They are all perfect scientists... They ask questions, they drive their parents crazy with Why? Why? Why? Then somehow they go to school and the school system crushes their curiosity, converts them to timidity and fear of science."

–Leon Lederman (Nobel Prize in Physics, 1988)

I had to end with the String theory, as we all understand is claimed to be the theory of everything, but naming this chapter, I didn't think about string theory but mostly on theories that may be the next few steps to inspire the journey ahead for students like me, who are keen on learning about astrophysics. But string theory is called the Theory of Everything because it consists of the standard model, which contains the basic particles that makeup almost everything around us, which explains the world around us, and you may never know. Maybe in the future we could find a particle that is so dense or so tiny that makes up 90% of objects and may even be added to the standard model. There is much more to learn about our universe; we are still standing at the beginning of our knowledge of the universe.

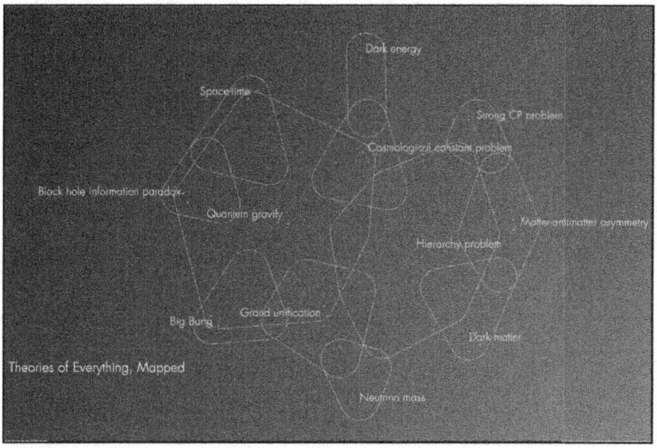

Photo Credit: Emily Fuhrman for Quanta Magazine, with text by Natalie Wolchover and art direction by Olena Shmahalo.

Source: https://www.quantamagazine.org/theories-of-everything-mapped-20150803/

❄ ❄ ❄

6

Climate Change

Climate change is still a major problem around the world, and has to be dealt with somehow. Our knowledge and understanding of our earth being in danger, from global warming, and rising temperatures, has come from many different fields. One field that may have contributed the most might be astronomy or astrophysics, as the study of the planets has helped us understand our planet compared to others, in terms of atmosphere, shape, gravity, temperature, and more. Thanks to astronomy, we have gained probably most of our knowledge on climate change and our environment, but what inventions have we created to help us discover deep inside our earth, and also beyond our earth? A recent theory that has been done by a student at Neils Bohr Institute, named Albert Sneppen, who researches the field of black holes and stars. Albert has started researching how global warming can be linked to the local weather, by looking at the distribution of light during the night sky, to analyze the temperature fluctuations and variations of the earth's surface, and maybe even compared to other planets, as the distribution of light all depends on the light and heat coming from the sun. This means that since the earth is very much close to the sun, the temperature will be an average climate. However, due to all the global warming and pollution being done, we are making the earth's surface much hotter than it is supposed to be. Sneppen's astrophysics approach is specifically designed to analyze this phenomenon known as the "Big Bang Afterglow". The electromagnetic radiation left over from the early universe is known as the "Big Bang Afterglow" or "Cosmic Microwave Background Radiation." This electromagnetic radiation, which is present everywhere in the universe, is a significant store of information, from the beginning of the universe. It is noteworthy that Sneppen discovered during this that the distribution of heat at the scales of the universe and the Earth together.

Photo Credit: Climate Change by TheDigitalArtist. Pixabay

Source: https://gnosisnetwork.org/blog/the-relationship-between-space-debris-and-climate-change/

Now, let's delve into a recent study that sheds light on the risky effects the sun can have on Earth, demonstrating the potential dangers it poses and the role astronomy can play in making ground-breaking discoveries to address climate change risk.

When Star flares occur many specific conditions may influence the magnetic field which can affect the flow of electrons in between the Sun and Earth. In technical terms, we can call these conditions low Planetary Index (K_p) which is a measurement that shows a planets stability and health, in this case our Earth. The flow of electrons weakening in technical terms can be called a low Electron flux (E-flux). The E-flux variation will, in turn, stimulate changes in the production of ionospheric currents. Ionospheric currents, referred to as flows of charged particles or ions in the ionosphere located in our Earth's atmosphere, will be influenced by this variation. These currents are important for electromagnetic variations such as how radio waves are produced. Ionospheric currents are produced by geomagnetic storms originating from the Sun-Earth environment. Geomagnetic storms are caused by the sun and affect our Earth's atmosphere, creating lights called auroras in our sky. These storms also involve the presence of ions. Can ionospheric current variations have an influence on our earth's atmospheric temperature? On 25th December 2004, hailstorm and snowstorm were reported in large areas in the northern hemisphere, while in the tropics a sudden drop of temperatures has led to foggy and smoggy conditions. This temperature variation can differ in different parts of the earth as a possible effect of the solar flare would be dependent on where the geomagnetic storms can occur which we can locate using geomagnetic coordinates. Sun-Earth environment K_p (planetary index), proton flux and electron flux show changes in behaviour. Sudden changes in these parameters may influence the environment of the earth abruptly. If an E-flux rise is responsible for global warming, then an E-flux lowering may lead to snowfall. On the 22nd of December 2004, a sudden fall in the electron flux was recorded by the SOHO satellite. Widespread snowfall was recorded in United Kingdom on the 25th of December 2005 (Aberdeen, London, Birmingham, Manchester, Cardiff, Belfast, Crosby, Woodford – Source: Meteorological Office, U.K., and BBC Weather News.). A subsequent rise of the E-flux normalised the condition. The star flare might have influenced the E-flux and thus caused a snowfall on 25th December 2004. Similar observations were noticed in other parts of the world also. Widespread snowfall was recorded in other parts of the world on the 25th of December 2004 and further on the 23rd of February 2005. Houghton, MI, Mauna Loa at Hawaii, Boston, and New York in the U.S.A. received very high snowfall; Tehran in Iran also received snowfall; Queensland in Australia experienced a cyclone with cold wave; Jammu and Kashmir, Shimla, India experienced a cold wave and received snowfall (Source: Meteorological Office, U.K., BBC Weather News, UK, and NOAA, USA). These weather conditions were irregular and were accompanied by low K_p index and low E-flux conditions. Researchers and scientists suggest that regular monitoring of Star flares and their influence on the Sun-Earth environment may lead to more accurate weather prediction.

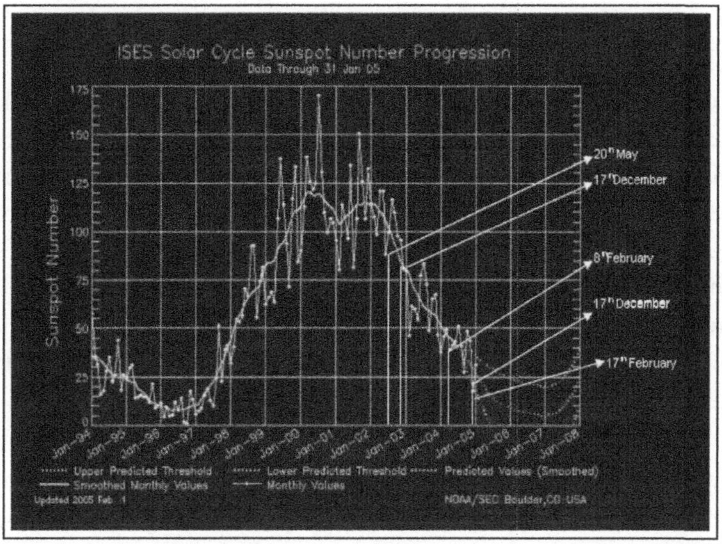

Influence of V838 Monoceras star flares (cosmic rays) on sunspot numbers. 20th May 2002, 17th December 2002, 8th February 2004, 17th December 2004 and 17th February 2005 show low sunspot numbers and higher cosmic ray (identified by cosmic ray detector). Courtesy NASA.

Saumitra Mukherjee, a geologist and remote sensing expert at Jawaharlal Nehru University (JNU) in New Delhi, India, shares a forward-looking perspective on future research in his book, "Sun-Earth-Cosmic Connection," published by LAP on 15th February 2011. Within the realm of cutting-edge interdisciplinary studies in contemporary science, the book emphasizes the Sun-Earth-Cosmic connection as an evolving technology designed to understand the dynamic processes in the Sun that influence the thermosphere, ionosphere, atmosphere, hydrosphere, biosphere, and the entire geosphere. Author posits that the integration of extra-terrestrial remote sensing observations and Earth-based sensors, such as cosmic ray detectors, harbours the potential to forecast various phenomena, including satellite lifespan in space, earthquakes, snowfall, rainfall, tornadoes, droughts, and epidemics within the biological system. This technology also holds the promise of contributing to the exploration of human dimensions in global environmental change by conducting comprehensive analyses of both human activities and natural changes.

Now back to inventions and ideas that astrophysics has developed to help us understand climate change and why understanding astronomy and astrophysics is important for us to learn more about climate change. Many organisations like NASA are helping to open up research on climate change. Here is an example. Observing the Climate of the Earth from the Astronomy Perspective. The study of the planet's changing climate is a priority for NASA. Their understanding of how the interconnected systems of our planet interact is being aided by the agency's observations of the planet from space, the air, and the ground. The organisation runs a significant program for climate research.

Moving on, one of the most interesting concerns posed by astronomers is whether or not we are alone in the universe. Why is a planet able to support life? From the Viking Mars Lander experiments, in the 1970s, through the first Venus missions, the desire to learn if there is extra-terrestrial life has been at the heart of numerous missions. Our improved knowledge of Mars and Venus, after numerous missions highlights Earth's special circumstances: the presence of steady conditions on a planet for billions of years, required to support life. But, why is astronomy important to understand climate change? Do we need to teach students about climate change? Do we need to leave it to the "experts" instead? The good news is that astronomers are ideally suited to teach climate change to students. The topics we teach in climate change are connected with science and theory of climate change, mostly. A typical technique to discuss climate change is, for instance, by discussing the climate history of the terrestrial planets. A most widespread misconception is that natural variability, not human action, can explain climate change, this subject is especially crucial. Another involves solar variability. Astrobiology and exoplanets are additional pathways and subjects. We use infrared telescopes in space, and on mountaintops, to reach above the water vapor that traps light, thus even topics like the nature of light and telescopes are related. We teach climate change through the words of astrophysics and astronomy, without even knowing it ourselves. There can be a lot of connections between the two. Who knows, maybe we will one day be able to save our planet through the eyes of astronomy.

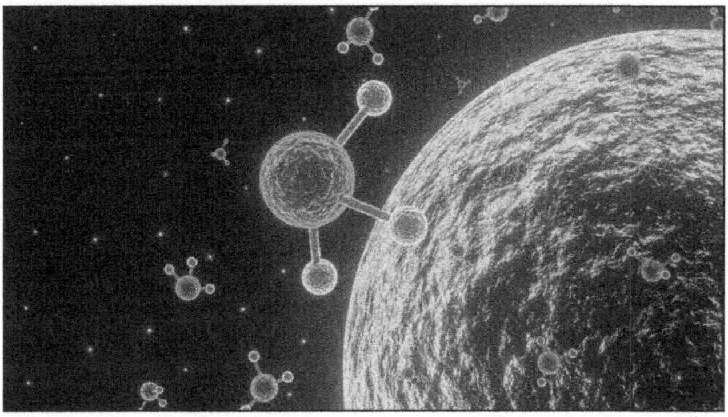

(to show life on other planets)

Photo Credit: Shutterstock

Source: https://newsroom.unsw.edu.au/news/science-tech/scientists-shed-more-light-molecules-linked-life-other-planets

❄ ❄ ❄

7

Future of Space Missions

Our physical space exploration was first initiated in the late 1900s, when we sent the first man into space, later, the first man to the moon, and then the first satellite on Mars, and so on. These explorations have helped us unlock scientific knowledge about our galaxy, and beyond. They have also helped us understand what may lie beyond our vast universe. However, we still don't know 97% of what our universe consists of. The use of technology has enabled us greatly to understand the world beyond us and enabled us to dream and innovate. But what new technologies could we create to help us learn more and more about our universe?

THE FIRST MAN IN SPACE

Gagarin, who was raised on a farm, as the son of a carpenter, earned his diploma in 1951, from a school close to Moscow. At the industrial college in Saratov, he continued his education, and took a flying course. After completing this training, he enrolled in the Soviet Air Force cadet school, in Orenburg, where he eventually graduated, in 1957. The Soviet spacecraft Vostok, which launched the first person into orbit, was used for the mission. Yury A. Gagarin was aboard the Vostok 1 spacecraft, which was launched on April 12th, 1961, and completed one orbit of the planet for the first time in human history.

Photo Credit: NASA

Source: https://www.nasa.gov/image-article/april-1961-first-human-entered-space/

FIRST-EVER SPACE MISSION

The first ever space mission without people involved was Sputnik I, the first man-made satellite, which was roughly the size of a beach ball, which was safely launched by the Soviet Union, on October 4th, 1957. On its elliptical course, it orbited Earth in around 98 minutes. New advancements in politics, the military, technology, and science were brought about by that launch. Despite being a single event, the Sputnik launch was the beginning of the space age.

Hubble Telescope

The significant telescope that has been in orbit for a while, is called the Hubble Space Telescope. It entered orbit on April 24th, 1990, thanks to the space shuttle Discovery. Hubble makes a 535 km orbit around the Earth. It weighs the equivalent of two adult elephants, and is the length of a huge school bus. Considering that Hubble moves at a speed of around 5 miles per second, it would take 10 minutes to travel from the eastern to the western coast of the United States. Astronomical objects including planets, stars, and galaxies are captured in clear images, by Hubble. One million or more observations have been made by Hubble, so far. Images of galaxies billions of light years away, comet fragments falling into Jupiter's atmosphere, and the birth and death of stars are among the details captured. The Hubble telescope uses ultraviolet rays and light to capture its mesmerizing pictures. Recently, NASA has come up with the James Webb Space Telescope (JWST) which has some new technological advancement, which we will discuss in a few pages.

Photo Credit: NASA/ Getty Images

***Sources*:** https://www.japantimes.co.jp/opinion/2023/06/29/commentary/world-commentary/hubble-telescope/

James Webb Space Telescope (JWST)

NASA's James Webb Space Telescope (JWST) is an infrared space observatory that was launched on Dec 25th, 2021, from the European Space Agency's launch site, at Kourou, in French Guiana. The $10 billion James Webb Space Telescope will help us uncover the depths of the history of the universe, from the Big Bang to our expanding universe, and more. It is one of NASA's greatest accomplishments, this huge space instrument that includes the likes of the Hubble Space Telescope, and that goes vastly deep into our universe. It took just 30 days for the James Webb Space Telescope to travel nearly a million miles (1.5 million kilometers) to its permanent destination Lagrange Point 2 — a gravitationally stable location in space, and an area for different sites to settle in, there are up to five Larganges, which show a different aspect and point of view than over positions. L2 is a spot in space, near Earth, that lies opposite the sun, and this orbit will allow the telescope to stay in line with Earth, as it orbits the sun. It has been a popular spot for several other space telescopes, as well. The powerful James Webb Space Telescope will take astonishing photos of the celestial objects, from when the universe started, as the light of those astronomical objects takes a very long to reach us. It's like travelling into the past. Luckily for astronomers, the Hubble Space Telescope remains in good health and the two telescopes will probably work together for JWST's first years to come, as the two are very likely similar to each other, though JWST is more technologically advanced. JWST will also look at exoplanets, previously identified by the Kepler Space Telescope. Thanks to a successful and neat launch, NASA announced that the JWST should have enough fuel to more than double its mission life expectancy of 10 years, to a whopping 20 years! Since, its launch, the James Webb Space Telescope's accomplishments just keep on coming and overflowing us with more knowledge, to uncover the depths and far beyond our universe.

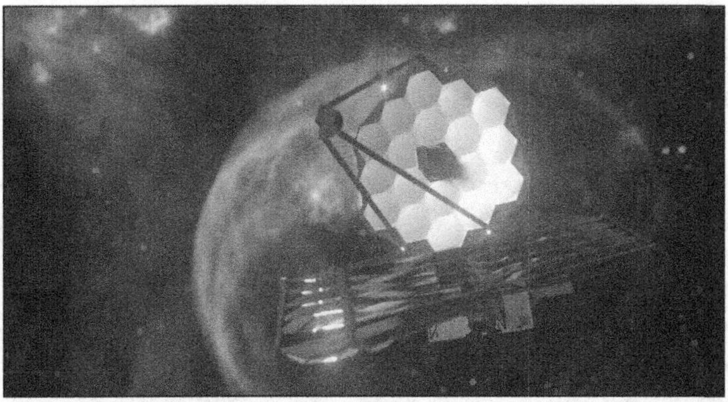

Photo Credit: Getty Images/iStockphoto

Source: https://nypost.com/2023/11/24/lifestyle/james-webb-telescope-discovers-new-galaxies/

Hope Probe Mars Mission

The UAE has sent their first ever satellite to Mars, to orbit around Mars for a few years, in order to collect some information about Mars's atmosphere. This would then be compared to the

Earth's atmosphere, and from that comparison we could assess how climate change is affecting our planet earth. It was launched on July 20th, 2022, at exactly 1:58 AM, GST time. It finally got the probe into Mars's orbit, on February 9th, 2021. This mission was directed, executed, and controlled by Her Excellency Sarah Al Amiri, who is the Minister of Space Science and Technological advancements. I had the privilege to meet Her Excellency in March 2021, as I had written an open letter regarding her amazing leadership towards the mission, and successful journey to the planet orbit; she instantly recognized my letter and invited me to her office to discuss how they had gone about the mission, and was also there to ask any questions I had regarding Mars and the UAE Hope probe. The UAE Hope probe is the first probe to provide a complete and clear picture of Mars's atmosphere, its layers and its temperature variations. There will be many other missions from the Middle Eastern area, especially the UAE, since there are many technological advances happening in Saudi Arabia. From Mars's data, we can figure about its habitability and if the planet can save the human race in the future.

Photo Credit: Khaleej Times

Source: https://www.khaleejtimes.com/space/uae-mars-mission-hope-probe-reveals-new-data-about-red-planet

CHANDRAYAAN 1 AND 2

Since it is a natural satellite of the Earth, the Moon is the nearest celestial body. We would be able to understand the history of the solar system and of Earth, by studying the evolution of the moon. The moon has been the subject and goal of several space missions, in India. With the launch of Chandrayaan-1 and Chandrayaan-2, the Indian Space Agency is not far behind others, as they have done quite a detailed research on the moon, and on other planets. Some water remnants on the moon were found by Chandrayaan-1. Chandrayaan-1 also discovered water ice in the Moon's North Pole. Magnesium, aluminum, and silicon were some of the minerals found on the Moon's surface, by Chandrayaan-1. The mission's worldwide imaging and the pictures captured by Chandrayaan-1 of the moon were another aspect of its success. On the other side, a soft landing on the moon and the deployment of a rover on the lunar surface were two of Chandrayaan-2's expanded goals, but they couldn't be accomplished. ISRO will undoubtedly be able to do this in the future, as many new technological advancements have been made that can help ISRO successfully achieve the mission, with Chandrayaan-2. In addition, to learning about the history of the solar system and the earth, we must first understand and investigate the evolution and revolutions of the moon and other planet's moons; let us first start with our natural satellite, the moon.

Photo Credit: rediff.com

Source: https://m.rediff.com/news/interview/why-chandrayaan-was-delayed-by-13-years/20190916.htm

CONCLUSION

The majority of engineers and scientists concur that this is a very exciting and hectic time to work in the space business. Over the Earth's atmosphere itself, several unprecedented things are occurring. Today, commercial companies can arrange for tourists to take short space flights into space; people are producing spacecraft for NASA missions, and a robotic chopper is currently roving Mars. Today's space travel has been transformed and advanced with the equipped technological advances, as we enter a revolutionary period. The moon is a goal for both NASA, private businesses and industries in the space sector. To start constructing a base camp on the moon, humans are expected to land there in 2025, as part of NASA's Artemis mission to inhabit the moon, and create new life there. This extended human presence on the lunar surface will aid NASA in preparing for longer-duration and farther-reaching human space exploration missions, including a potential people crewed voyage to Mars, and improve space exploration in our human race. We need to advance the new space age and help us explore the depths of our universe and beyond.

Photo Credit: Artist impression of the Tianwen-1 Mission/Nature Magazine

Source: https://www.nature.com/articles/s41550-020-1148-6/figures/1

❄ ❄ ❄

Made in the USA
Monee, IL
08 November 2026

68174120R00069